Fieldwork

The Life and Mind of a Biologist

Raymond Clarke, PhD

Copyright © 2024 Raymond Clarke
All rights reserved. No part of this publication may be reproduced, distributed, or transmitted in any form or by any means, including photocopying, recording, or other electronic or mechanical methods, without the prior written permission of the publisher, except as permitted by U.S. copyright law.

ISBN: 979-8-218-51137-1

Cover photo by Brad Gemmell

To

Serena and Justin

Karen and Brigid

Without them it would all be meaningless.

The most beautiful thing we can experience is the mysterious. It is the source of all true art and science. He to whom the emotion is a stranger, who can no longer pause to wonder and stand wrapped in awe, is as good as dead—his eyes are closed.

Albert Einstein

Somewhere, something incredible is waiting to be known.

Carl Sagan

Table of Contents

Prologue 2
Maps 3
Roots 5
    Starting with Fish 10
    Fishing 13
    Hooked 14
Becoming a Biologist 16
    Summers with Miss Patton 19
    Early School Days 24
    Undergraduate Studies 27
    A Close Call 28
    Choosing a Path 29
Encountering Reefs 34
Learning to Dive (In Spite of Myself) 37
    The Brain and the Gut 41
Back on Solid Ground 45
The Nature of Science 48
    More Play 50
Fieldwork 53
    Sometimes You Have to Get Creative 54
    Sometimes You Get Hurt 60
    Boating for Science 63
    Sometimes You Get to Live a Dream 67
    Research in Hydrolab 75
    Joys and Sometimes Dangers of Hydrolab 77
My Hippie Summer 81
Sarah Lawrence College 87
My Fascination 91
    Ideal Model Organisms 92
    Habitat and Behaviors 95
    Where to Begin with Blennies? 96
    Resource Partitioning 98
    Collaboration is Key 106
Harsh Environments 113
    Winters in Montreal 114

Becoming a Father	119
Halcyon Environments	121
The West Indies Lab	128
My Reckoning	133
Living and Dying	135
Animals and Science	136
Octopus Stories	142
A Soft Spot	146
Cats	148
Discovery of a Species	152
Communicating Science	155
Size Does Matter	159
Science is Conservative (But Not How You Might Think)	162
My Father's Legacy	166
His Demise	171
Up in Smoke	174
From the Ashes	178
Shifting Baselines	181
The "Good" Old Days	183
Loss, and More Loss	186
Winding Down	189
Once and Forever a Biologist	193
Nature, Nurture, and the Need to Communicate	198
Nurturing Nature	201
On the Otter Hand	203
Using Scavengers	207
Evolution and My Mind	211
Evolution Isn't Always Efficient	215
The Meaning of Life, A Meaningful Life	218
"There but for the Grace of God Go I"	221
Bringing It Home	222
A Final Thought	226
Epilogue	227
Acknowledgements	232
Endnotes	233

Prologue

I had scuba-dived on that coral reef many times before and was familiar with the eye-catching abundance of fishes: blue-headed wrasses, ocean surgeonfish, Nassau groupers, stoplight parrotfish, and a hundred others. They slowly cruised about as individuals, pairs, and small schools, constantly in motion, forming a halo over the corals, sponges, and algae that encrust the reef. But on this day in 1979, I peeked under the halo by placing my masked face within inches of the surface, and not only did I see the tiny shrimp, crabs, and worms that I had expected, but also fish! Teeny fish, only an inch long. These were not baby fish; they were adults. And they were not swimming about like other fish. These guys lived in holes in the stony coral surface—very unfishlike!

The face of a coral reef is pocked with cavities made by an assortment of creatures that burrow into the stony material—mussels, worms, barnacles, even sponges. It's a Swiss cheese. And many of these holes are occupied by little fish, bodies hidden within, heads projecting out, looking, always looking, for predators to avoid and food to eat. The more I scanned, the more I saw. These little hole-dwellers were the most abundant fish on the reef, and I had been totally unaware of them. My perspective completely changed. I now envisioned the fish community on a reef as lots of big fish openly swimming about and even more tiny fish living on the surface, or rather in the surface—cryptic fish. Time to recalibrate my perception. That recalibration changed the direction of my life.

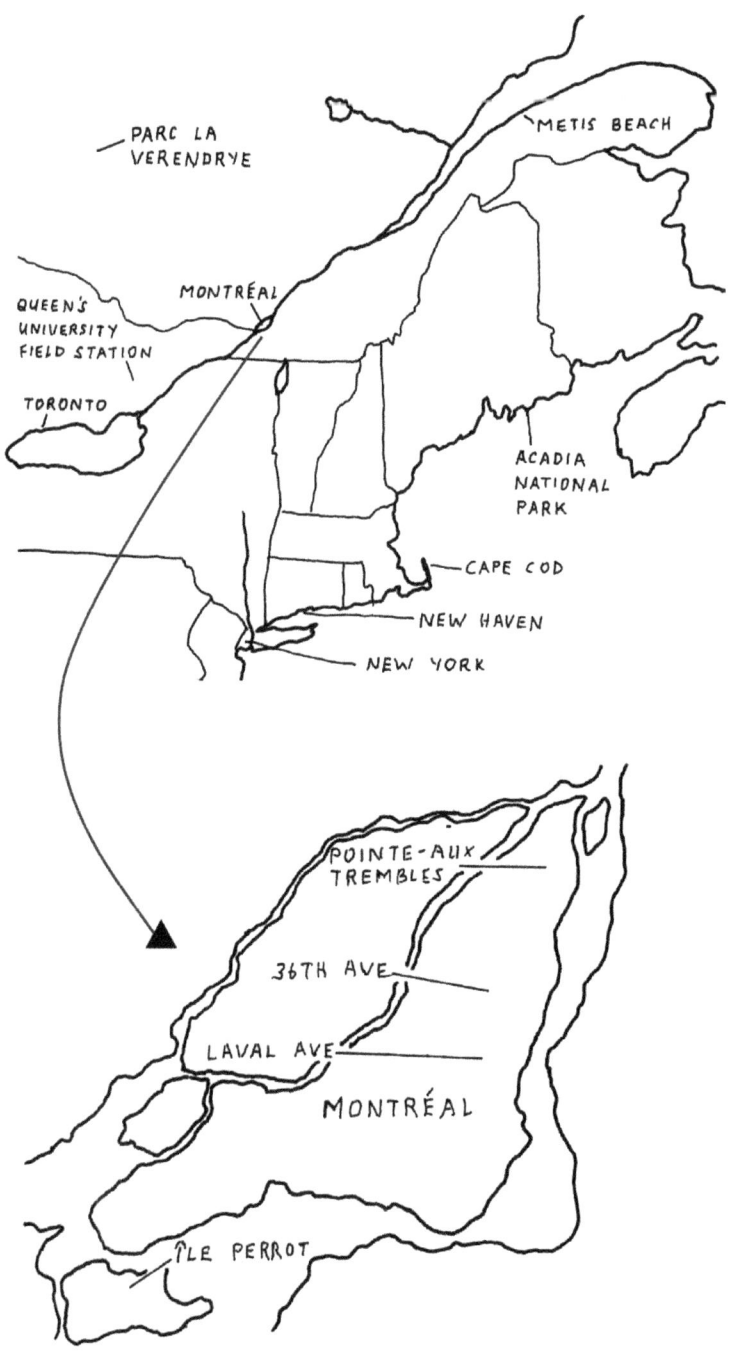

Roots

It's hard to separate influences, but my father was male AND English, and my mother was female AND Ukrainian. Those combined features gave me a cool, distant dad and a warm, loving mum—tight hugs from her, a handshake from him. My dad did not speak much of his past, but as far as I can determine, his parents were stern, and they led a severe life. Getting an orange at Christmas was a real treat for him. He was a man of his era, which means he did not show his emotions. I'm sure he loved me but did not know how to engage with children, or maybe he simply felt that treating me in a tough manner was a way to prepare me for the world. In any case, I worshipped him and followed him everywhere. Relatives would frequently say: "There goes Cyril's tail." I never spoke with my parents about how they raised me, even in their later years. I did resolve to be more outwardly loving than

Nanny with daughters in 1934. Peggy on left and Stella (my mum) on right. Photo from entry document to US.

Family circa 1953. Author center left, Wendy bottom left, Dad top left and Mum second left center row. Grandparents in center and Peggy with her family on right. Photo from family collection.

my dad and managed to do that, but not completely. I am still somewhat distant. It's not easy for me to pick up the phone and just start talking with them. I don't like that about myself, but I find it difficult to change.

Stella, my mum, immigrated to Montreal, Canada, as a little girl in 1934, along with her sister and mother. Her father had come earlier to get established. They first landed at Ellis Island in New York, and she recalls staring at a man in the waiting room as he ate a strange alien fruit, an orange. So much was new to them. In 1958, my parents and grandparents bought a house together where we lived, an extended family of six, including my younger sister Wendy. My grandparents never learned much English, so Mum spoke to them in Ukrainian and to Dad and us kids in English. My grandfather, a kind, gentle man who worked as a night watchman at a bakery, sometimes brought home treats—day-old doughnuts. My grandmother was distant and suffered from headaches. At these times, she would slice raw potatoes and place them around her head, held in place by a knotted cloth. We kids were thankful that Mum did not apply Nanny's remedies to us.

Mum, a consummate nurturer, insisted I wear a hat when going out in winter, but hats were not trendy, and I invariably took it off halfway up the block. I came home for lunch every day when I was in elementary school. Once in a while, Mum made crêpes; we called them French pancakes. On opening the front door and smelling them, I would jump up and down, yelling: "Yippeee!" That brought a happy smile to her face. I smothered the crêpes in Lyle's golden sugar syrup and greedily gobbled them up. I still remember that syrup. The can had a drawing of a recumbent lion with a swarm of bees flying around it. Beneath the drawing was the motto: "Out of the strong came forth sweetness."

"Out of the strong came forth sweetness." That's a perfect description of my mother. She was very energetic—constantly cooking, cleaning, shopping, more cleaning. Her dentist, based on her worn teeth from vigorous brushing, said that she must be a fastidious housekeeper. And her neighbors were impressed that she mowed the lawn herself well into her seventies after my father died.

One of my first memories is my standing on a chair beside her, peering into the pot as she was cooking porridge on the kitchen stove. I must have been between three and four years old. I asked her what those bubbles were, and she said they were little bears jumping out of the porridge—at least I think that's what she said. In the warm kitchen, I was happy, and my world was secure—my mother was there. And she was there all the time. When I visited her with a girlfriend after divorcing Karen, my first wife, we two went out to dinner and then stopped at a tavern for a beer. We got home at two in the morning, and Mum

Mum and Dad, 1945. Photo by Meyers Studio.

was up waiting and was really angry. She had imagined all kinds of awful things had happened to us. I was fifty-one years old at the time.

My Mum and Dad met in Montreal during World War II. He was a radio operator (Morse code), and she inspected airplane parts in a factory supplying the military. I have a photo of them taken just after the war. He is proudly wearing his RAF uniform; she is looking happy, bright-eyed. Mum had dark brown wavy hair that stayed dark until the end. Dad said that when they first met, she had such a small waist that he could put his hands around it with the fingertips touching. She may have had a small waist, but he also had large hands. Come to think of it, I was probably conceived about the time that photo was taken.

As I was growing up, I saw my parents getting along fine. There was an occasional argument, nothing physical, with some yelling, but that was rare. My dad would speak in a dismissive manner to her, and she would generally not respond. You might think Dad was the dominant one, but quietly, behind the scenes, it was Mum who ran the show. She was stronger than her demeanor suggested. For example, when Max, our beagle, was getting old and incontinent, she took him to the vet and had him "put down." Dad loved that dog, but Mum said it had to be done and did it without telling anyone beforehand. We got home, and Max wasn't there. Tough. Back then, nobody would consider putting a diaper on a failing dog as many people do now.

Dad loved being outdoors, and that is why he chose to be a bricklayer. He had a seventh-grade education, and Mum went as far as the ninth grade. My parents wanted me to read and encouraged me to do well in school. There were few books in the house other than a twenty-volume children's encyclopedia, *The Book of Knowledge*,[1] that my mum had purchased from a door-to-door salesman when I was nine years old. The set filled two shelves on the small wood bookcase that came with the package deal they paid in monthly installments. A little later, three other books joined the encyclopedia. I had copies of Rachel Carson's *The Sea Around Us*,[2] *The Red Book of Birds*,[3] and *The Blue Book of Birds*.[4] The bird books were small format, about four by five inches, and each page was devoted to one species. They were part of a four-volume set, but I did not have the *Yellow Book of Birds*[5] or the *Green Books of Birds*.[6] Sitting for hours studying those two books, I learned to identify many birds, but I missed out on half the diversity (warblers, wrens, swallows, thrushes, and others). To this day, I still have greater knowledge of the birds in the red book and blue book. Mum's dream was for me to go to the local university, McGill. I did. Later, when I went on to graduate school, my parents had no idea what that meant.

Starting with Fish

A shy kid, quite self-sufficient, I could happily spend hours on my own exploring parks, empty fields, and regenerating forests on the edge of town. My dad loved Nature in his own way, and he exposed me to animals and showed me the wonderful richness of the natural world. He had an aquarium from the earliest days I remember. I can still envision it: a large glass box with a steel frame, light reaching down from above illuminating the bright green masses of aquatic plants, some with long strap-like leaves that reached to the surface and others with broad lanceolate leaves and red stems. A variety of colorful fish swam about, some hanging out among the plants near the bottom, others moving freely in the open upper levels. I got excited every time we took the bus to the pet store to buy a new fish. The rows upon rows of tanks with a seemingly endless variety of fishes overwhelmed me. There was so much to choose from, but we would only buy one or two individuals. Plastic bags did not exist back then, and the fish were placed in repurposed glass jars for the trip back home. I didn't think of it at the time, but now I wonder where the shop owner collected so many empty jars, one for each purchase.

Over time, a fluffy brown layer of organic waste produced by the fish covered the sand at the bottom of the tank—time to clean it out. Dad caught the fish in a net and put them in a bucket of water. He placed the end of a piece of rubber tubing in the tank and sucked on the other end to start a siphon draining the water into another bucket below the tank. Sometimes, he didn't time it right and got a mouthful of water. He jerked his head back and spit the water into the bucket. He removed the plants and took out the sand to rinse clean. It had a strong, earthy smell, the rich, pleasant smell of abundant life. He returned the sand to the aquarium, reached down, holding each plant by the roots and sliding it through the sand to plant it. Then it was time to place a sheet of waxed paper over the plants to prevent their being dislodged as he poured water in from a hose. I watched all this with great interest.

I absorbed my dad's love of Nature and learned so much from him. He seldom instructed me; he just quietly did his thing while I watched and absorbed. We definitely had a bond. Being close and observing created a warm feeling in me and probably in him, too.

He fed the fish daily with dried granules of food from a small can purchased from the pet store. He also occasionally fed them white worms. These were a little thicker than a hair and about the length of a fingernail. After becoming a biologist, I learned that these worms belonged to the family Enchytraeidae and that they normally lived in moist soil under the decaying leaf litter of deciduous forests. I occasionally saw them when I turned over leaf litter, looking for interesting things. Dad cultured these worms in a small wooden box filled with soil. He placed a piece of bread soaked in milk on top of the soil in the center of the box. Then he covered it with a piece of broken window glass and several water-soaked sheets of newspaper to keep it dark and moist. At feeding time, he lifted off the newspaper, and there against the glass was the smooth white bread turned into a paste. Thick masses of white worms encircled the edges of the bread as they fed on the mush. Dad lifted off the glass, and some worms came up, stuck in the wet underside of the sheet. But most remained on the soil, and he pinched off a mass of writhing worms to drop in the fish tank.

The fish swam about excitedly, and each dashed in to take a bite of the worms, except the angelfish. These were larger and more sedate than the others, and they parked by the mass of worms, taking bite after bite. Angelfish were the most exotic aquarium fish—silver discs with long flowing fins above and below, all broken by a few black vertical bars. Most of the aquarium fish were bred in captivity but originated in the Amazon River basin. I fantasized about being in the Amazon, the most fascinating of places, with pink dolphins, giant otters, tapirs, and piranhas. Sixty years later, I saw all of these on a trip to the Peruvian Amazon with my friend Dan, an avid birder and frequent traveler. We were staying on the Tahuayo River, a small tributary of the Amazon. One night we took a boat ride, poking around using our headlamps to look along the shore for animals and in the water for fish. A light rain started, but we simply ignored it in the warmth of the rainforest. We drifted into a little side channel choked with vegetation, and our naturalist/boatman/guide reached his hand into the water and lifted out an angelfish. It lay quietly, flat on his hand, silver with black vertical bars and elongated vertical fins, just like in my dad's aquarium, until he gently immersed his hand, and it scooted off.

Author with lionfish, Andros, Bahamas, 2008. Photo by N. Nakamura.

My dad had a book of aquarium fishes with a beat-up gray paper cover. Each page had a black and white drawing on the top half and a description on the bottom half. These were all freshwater fishes, except for the last page, which displayed a drawing of a lionfish. Its narrow head, white and filled with thin black stripes, was facing directly at me, its large mouth downturned as if in a frown. On either side of the head, its extensive pectoral fins fanned out, a membrane between the supporting rays except at the tips, which extended well beyond the membrane with a downward hook. If angelfish were exotic, this fish was ultra-exotic. It looked like something from another planet, and I could only imagine how it appeared in life. In fact, it was a marine fish from the southwest Pacific.

Many years later, I saw several lionfish while diving in the Bahamas. They had been introduced into the western Atlantic, whether by accident or on purpose, nobody knows. We do know that Caribbean fish have no experience with this unusual predator and its unique hunting method. It uses its gigantic pectoral fins to scare little fish into hiding where it can slurp them up. This fish is now seen as a major source of mortality for small fishes, both adults of small species and the young of large species. It is now literally an exotic, and like many introduced species, it is wreaking havoc with the native community. Sixty years after dreaming about that drawing, I gave a seminar at Sarah Lawrence College on the spread and impact of lionfish in the western Atlantic. Now, there are recipes for lionfish on many websites, and some restaurants are offering lionfish on the menu. We have learned the lesson of history: if you want to decimate a species, make it commercially valuable and begin harvesting it.

Fishing

Seven years old, sitting in the bow as my father rowed our shallop, a simple wooden boat that he had rented for the afternoon, I looked into the water, got really nervous, and said: "Stop. The water's too deep. It's over our heads." He did not stop. We were fishing at Île Perrot, an island off the west end of Montreal. The water was quite clear back then, and we could see the bottom, covered in tall feather-boa-like plants among which the fish foraged. We put worms on hooks and caught bullheads and yellow perch. When we got home, he filled the bathtub with water and dumped the fish in, and they swam around. Many turned upside down at the surface as they died, but the bullheads were really tough and did fine. I remember his cleaning the fish and using a pair of pliers to pull the tough skin off the bullheads. He would open the stomachs to see what they had eaten, and scads of snail shells tumbled out of the perches' stomachs. After my conducting research for several years, Dad asked if I was still studying fish. He seemed to think that, at some point, I would know all there was to know about fish. He also asked me when I was going to study "real fish." To him, the fish in the ocean were weird creatures, and freshwater fish, trout, pike, and the like were real fish.

Hooked

Before my father ever took me fishing with him, he went on his own, and I saw him return with his catch. I indirectly knew about fishing, and I'd seen his rod and reel—and the hooks.

In the late afternoon of a bright, sunny weekday in 1952, Laval Street in downtown Montreal was quiet. All the fathers were at work, and the mothers were inside cleaning house and preparing dinner. Two six-year-old boys were playing on the sidewalk. They were make-believe fishing, and one boy (the fisherman) sat on the stoop holding a line while the other (the fish) was crawling on the sidewalk. The fisherman threw out his hook, and the "fish" grabbed it in his hand and pretended to struggle while he was hauled in. The boys were taking turns being the fish and the fisherman.

It was my turn to be the fish. I told my friend that when my dad catches a fish, the hook is in its mouth, and I proceeded to act like a real fish by putting the hook in my mouth and crawling around on the sidewalk. All was going well; fish and fisherman were playing their roles. My friend was holding the line that was loosely stretched across the sidewalk with me on the other end, the struggling fish with a hook in its mouth. A man came walking up the sidewalk, wearing a tie, loose jacket, baggy pants, and a fedora. He was aware that two boys were playing some game as he walked between them, scanning the evening newspaper, *The Montreal Star*. He snagged his foot on something but didn't fall. I felt a sudden sharp pain as the hook punctured my lip. Now I <u>was</u> hooked, blood flowing, and I was howling and yowling. My mother came running down the stairs, and heads popped out of windows up and down the street. The man dropped his newspaper and ran over to me, very upset. People gathered around, gawking. The man felt guilty and said: "I didn't see the string." I was shaking uncontrollably, and Mum comforted me in her arms.

She called a taxi, and when it arrived, she helped me in and we headed off to the doctor's office. We were frantic. I was still shaking, but the doctor, apparently having seen everything during his years of practice, calmly took a pair of surgical pliers, grabbed the shank of the hook, and extracted it with a deft twist. There were no stitches and no bandages; there's little that can be done to the inside of a lip. Arterioles contracted, blood clotted, and the bleeding stopped. Young tissue heals quickly.

When Dad got home from work that evening, he had to face my mother. It was not a pretty sight. "What were you thinking, giving him a fish hook to play with?"

"But I filed down the point and the barb so it would be safe. I never thought he would put it in his mouth."

My Dad did a stupid thing, and I did a stupid thing. It's fortunate he filed off the barb or there would have been a lot of tearing of tissue when the hook was extracted. Then the healing would not have gone so smoothly. As it is, I have no scar.

Becoming a Biologist

Until I turned eleven, we lived in a rented row house in downtown Montreal. The small backyard, enclosed by a rough solid wooden fence, consisted of bare earth and had a door opening on the lane that ran between the houses on Laval Avenue, where we lived, and Avenue de l'Hôtel-de-Ville. A gigantic cottonwood tree occupied a corner of the yard next door, and every spring it would release clouds of white fluff that drifted around and covered everything. In the backyard, I caught tent caterpillars and placed them in jars with nail holes in the lid for air exchange and oak leaves for food. I did not know why, but the caterpillars always died. I was totally unaware of both the very specific food requirements of caterpillars and of plant chemical defenses. When removing the jar lids, I could smell the heavy, pungent odor given off by the leaves, but I did not connect that with the caterpillars' death.

I used to walk several blocks to Mount Royal, the large park in the center of Montreal, and catch insects. One fall day, I had a large praying mantis in a shopping bag, and a mounted policeman stopped me, asking to see what I had. He was very stern and looked twenty feet tall way up on that horse. I held up the open bag above my head to show him. He looked in and, with a disgusted snort, waved me on my way. I have no idea what he thought I was up to, a seven-year-old boy in the woods with a shopping bag.

Once, I came back from a trip to the country with a bunch of little jet-black tadpoles in a gallon jar. I left the jar out sitting in the sun against a wall in the back yard, and the water turned bright green with algae. While still the size of a corn kernel, the tadpoles developed legs, lost their tails, and a little later, scores of tiny toadlets hopped out of that jar. What foreshadowing! This observation by my eight-year-old self was followed twenty years later by a published paper from my doctoral dissertation in which I showed, among other things, that toad tadpoles transformed at a smaller size than all other frog-like animals.[7]

In 1957, I moved with my family from the Laval Avenue house in central Montreal to Rosemont, a section in the northeast corner of the city. Our street, 36th Avenue, consisted entirely of identical semidetached red brick duplexes. The area was rather new. There were woods a short walk away and patches of undeveloped fields scattered throughout the sprawling, repeating streets lined with repeating houses.

Tall grasses covered one abandoned field a couple of blocks away. In late summer and fall, the grasses were so tall they hid us completely when we lay down. After school, I would meet there with some friends and sit, talk, and play word games. It's there that, in response to a dare, I kissed a girl for the first time. It wasn't a hot lingering movie screen kiss—a peck would be a better description. I didn't even like her that much. A couple of years later, that field became a supermarket and more tract houses.

Another field, a block away in the opposite direction, was more ragged, with patches of tall grass, short grass, and rocky outcrops. It was a perfect habitat for the meadowlarks that nested there. The outcrops contained loose pieces of sedimentary rock. Embedded in these stones were sinuous rows of disks less than a quarter inch wide and of varying length, many short but some more than three inches long. These black and shiny truncated strings stood out against the light gray flat powdery stone. They looked like sections of necklaces. I knew they were fossils, but only after entering college five years later did I understand that they were the stems of extinct sea lilies about 450 million years old. It took several more years before this plot was developed. It became a city park with lawns, paved paths, and a playground. In the winter, a section was flooded and became a skating rink surrounded by a four-foot-tall solid board fence for hockey. I skated there, as much on my ankles as on the blades, because Mum bought me skates a couple of sizes too big, expecting I would grow into them. I never did. She frequently purchased my clothes a bit too large. As she shared with my aunt: "They grow like weeds."

About fifteen blocks away and much larger than the others, the most interesting undeveloped area consisted of a young forest growing on abandoned agricultural land. As it was throughout the region, the land was flat, with no hills or valleys. The tree trunks were the thickness of my thigh. It was here that I hunted for snakes. They captivated me by their legless otherness, their ability to seamlessly fit their flexible, coiled bodies into the depressions in the ground. I would wander through the trees turning over rocks, and once in a while, it would be there: a snake loosely coiled, suddenly exposed in its hiding place. Every time this happened, I would freeze, crouched over the snake, my heart pounding—probably the snake's, too. There we would remain, momentarily immobilized until the snake suddenly made a dash for safety. That's when I would make my grab, usually successful, and end up with a writhing snake in my hand.

Two species were no more than a foot in length: the drab gray-brown DeKay's snake and the beautiful ringneck snake, with an elegant smooth charcoal gray back and a coral-red belly, the red also forming a ring just behind the head. Two species were about two feet long: the green snake, which, as its name implies, is a uniform bright lime green in color, and the garter snake, a gray-green snake with three yellow stripes running down its length. If it was a green snake, it might bite me. If it was a garter snake, it would probably release a foul-smelling musk on my hand. The other species were totally benign. I once caught a milk snake—the big prize—three feet long, light brown with irregular darker brown patches on its back.

Usually, I let the snakes go free, but a few times I brought one home to keep in an empty aquarium. I tried to give them an inviting environment and feed them. My mother hated this but tolerated it. Her dislike was appropriate because they invariably escaped after a day or two. I don't know their fates. We never saw one loose in the house. I assume they found their way out, but I doubt they survived among the lawns, driveways, and roads.

Wandering around those woods alone was usually a very pleasurable experience for me. But on one occasion, as I turned a corner along a trail, I came face-to-face with three boys about my age, one holding a BB gun. We stopped in our tracks, facing each other, and I felt like a snake whose rock shelter had just been lifted. The first boy fired the

BB gun, hitting me in the upper left thigh. I turned, terrified, and ran as fast as I could, ignoring the branches whipping my body as I passed. They didn't follow. Upon arriving home, I pulled down my pants to check myself and found a little blue mark where I'd been hit. It didn't really hurt much and disappeared in a few days. Why did they shoot at me? It may have been French-English animosity in Quebec at the time, or simply a boy with a gun—he'll find something to shoot!

More development, largely housing, displaced the whole forest several years later. In 1967, I moved to New Haven, Connecticut, for graduate school. On returning home for Christmas, I went to a mall to buy presents. That mall sat on the location where I used to catch snakes. Having been salted, the roads were black and wet, but snow covered the lawns, driveways, and other sterile surfaces. The only visible trees were isolated skinny saplings in some front yards.

Urban sprawl happens, but you don't have to like it. To developers, it's their livelihood. To Nature lovers, it's a disaster. I vividly recall a time, later in my life, when I was driving up North Highland Avenue in Ossining on a summer's day and came to the spot where Cedar Lane veers off to the right. Those two streets had bordered a little triangular remnant piece of forest. But this time as I approached, the trees were gone, and a big yellow bulldozer sat in the empty space, bathed in intense sun where before there had been deep shade. The nurturing soil, now ripped and ragged, had been torn up as trees were pushed over and bulldozer tracks cut their way through. I was shocked. My heart sank. It felt like a death. And what went up on that site? One more strip mall.

Summers with Miss Patton

Wendy did not come with us when we made that move to Rosemont because she was away for the summer. My sister's first-grade teacher had invited her to her country home in Metis Beach on the Gaspé Peninsula. Miss Patton was a cranky, elderly spinster, tall with a hunched back, a thick mass of unruly gray hair on her head, and single hairs sprouting from various places on her spectacled face. Known for being very strict, she punished misbehaving boys by grabbing them by the ear and walking them to the corner. But she truly did like children and every year asked a few to spend their summers in the country with her. Apparently, Wendy spoke about me incessantly until Miss Patton wrote my parents and invited me to join them. My parents sent me along, and I spent five glorious summers at Metis Beach.

Each trip started with a ride on the Ocean Limited, a train that ran from Montreal to Halifax. A gigantic diesel locomotive followed by a baggage car pulled the long string of passenger cars, trailed by a traditional red caboose. A porter in a white uniform attended each sleeping car. Soon after we left Central Station, around sunset, the porter would fold up each set of facing seats into an upper and lower berth. Each bed was separated from the central aisle by very heavy dark green drapes. It was a level of luxury I wasn't used to, but I enjoyed every minute of it. I remember trying to sleep in my excited state in that swaying berth with the clack-clack, clack-clack of the wheels on the track and the squealing of the brakes and explosion of released steam as we pulled into a station in the middle of the night. At each stop, I would peek through the curtains and see a brightly lit but empty platform. Those stations had a forlorn look in the night, giving me the same impression that I got many years later when I looked at Edward Hopper's painting *Nighthawks*. That artwork portrays a dark, empty street late at night, the only light streaming from inside a bright corner restaurant with three people sitting at the bar. To me, it portrays absolute stillness, just like those empty train stations. I've always been impressed by Hopper's ability to evoke emotional reactions with the clean simplicity of his images.

Miss Patton's father had been a prominent doctor in Montreal, and he left her with a very large property, including a three-story wooden house sitting on a rocky promontory on the seashore. The last of the line, aging and childless, she lived in the decaying house in which she had spent her vibrant childhood. She was much like Miss Havisham, a character in Dickens's *Great Expectations*.[8] Ever since she had been jilted at the altar, Miss Havisham wore her wedding dress and never left her ruined mansion, with the clocks stopped and the dining room table set for a wedding feast, the food untouched, shriveling as it decayed. This strong image has remained with me ever since I read the book in high school. Miss Patton was much like Miss Havisham, including a love tragedy in her past, though she wasn't jilted. Her fiancé had been killed fighting on the British side in the Boer War in South Africa (1899 to 1902). A souvenir of that war, a bayonet, lay on the table beside her couch. We assumed it was her fiancé's but did not really know.

The house had a wrap-around porch, a decrepit tennis court, and a brass PATTON nameplate on the front door that she had us polish weekly with Brasso. The downspouts led to big wooden rain barrels from which we drew soft water to wash our hair. I would occasionally remove the wooden cover from one and watch the aquatic insects swimming in the clear dark water—little twitchy mosquito larvae, larger twisty phantom midge larvae, and an occasional black beetle motoring around. The property encompassed a large rocky hill with a vertical 100-foot cliff facing the sea. We knew it as Mount Patton, but I heard that everyone else in the region called it Mount Misery. Apparently, a ship had foundered on a nearby reef, and several bodies had washed up at the foot of the cliff many years before.

Besides Wendy and me, two or three other favorite students would stay for the summer with Miss Patton. She would spend the day lying on her wicker couch, wearing her fiancé's old army jacket, with a drink in her hand, smoking cigarettes, and reading paperback novels. Being the oldest kid, it was me she sent once a week through the fields and down the highway to Mr. Meikel's motel to pick up "a package" for her. He would always hand me the heavy brown paper bag with an awkwardness that made me feel something surreptitious was going on. Miss Patton could be testy but could also be kind. She taught us a great deal. We cleaned the house and did a lot of the cooking under her supervision. The kitchen had a big wood-burning stove that we used for much of the cooking. We made toast by placing the bread directly on the iron surface of the stove, but only after conditioning it by rubbing it with the waxed paper in which our bread came packaged. Once, when a new girl arrived and asked where to make the toast, we pointed to the stove, and she dropped the bread into an open pot of steaming water. We laughed at her, and she felt terribly humiliated. A favorite treat we'd make was a heavy dark brown steamed pudding called Sailor's Duff. We topped it with a semi-sweet white sauce and loved it.

Those summers were an incredible gift to me from age eleven to fifteen. I spent endless hours alone, happily wandering through the woods and along the seashore. I walked barefoot much of the time, and my feet toughened to the point where I could run at full speed over the irregular vertical layers of shale on the beach. At low tide, I would head out onto the exposed rockweed-covered ocean floor, inspecting tide pools and finding sea anemones and beautiful pink frilly sea slugs. I overturned rocks and occasionally found a rock gunnel, a small eel-like fish, protecting its eggs. I made an aquarium out of a gallon jar and kept a little community in there. I recall watching a sea anemone releasing babies from the tips of its tentacles in that jar.

I also explored the surrounding spruce woods. Many of the trees had large patches of missing bark where porcupines had been feeding. I once came across a woodpecker's nest ten feet up in a dead white birch tree. The trunk was about eight inches in diameter, and although quite rotten, the tight bark was intact and kept it together. I saw the parents flying in and out of the hole, feeding their young, who chattered noisily whenever a parent arrived. I wanted to take care of those little birds, so I pushed the rotting tree rhythmically, and it swayed through ever-greater arcs until it came crashing down. I gathered up my tiny prizes and made a little nest for them in a cardboard box and fed them regularly. They died. Feeling awful and regretting my thoughtlessness, I buried them. I had no idea of the nutritional and other needs of baby birds. I still don't know the needs, but I do understand that they are specific, and I would never think about engaging with such hubris again.

An open patch of raspberry bushes about four feet high covered one section of the mountain. In shorts, I carefully picked my way through the thorny canes on a clear day with a hot sun overhead. I put a couple of berries into my pail, took a step or two, collected another couple of berries, took another step, and all hell broke loose! I had stepped on a yellow jacket nest, and a swarm of wasps boiled up around me. I took off as fast as I could, thorns tearing my bare legs, running right through the bushes until I no longer felt threatened. I stopped, my lungs heaving and my heart pounding. I looked up, and high over the water, I saw an osprey and heard its unique piercing cry. With my brain in a hyper-alert state, my memory of that instant is so vivid; it could have been this morning. I was proud of my swift reactions that resulted in only four stings, but they came at the cost of countless cuts and scratches on my bare legs.

I wasn't the only one with a traumatic encounter up on the mountain. Miss Patton's dog, Fido, had a run-in with a porcupine and arrived at the back door with a snout full of spines. There was nothing to do but remove them with a pair of pliers, and being the eldest, therefore the "officer in charge" or the "OC" as Miss Patton used to say, it was up to me. Now, Fido was the sweetest dog you could imagine, but porcupine spines have tiny backward-pointing barbs that tear the tissue when pulled out. I knew what I had to do because my father had shown me, in a way.

One sunny afternoon in the summer of 1952, I was playing with my cousins in an abandoned car rusting among the weeds behind their apartment in Pointe-aux-Trembles, a suburb to the east of Montreal. I was sitting on the roof with my hand grasping the edge of the door opening, and my cousin Larry was in the driver's seat. He slammed the door shut on my finger, and I let out a scream. Crying and in pain, I jumped down and, shaking like a leaf, ran into the apartment where the adults were chatting. They all gathered around me to inspect the bleeding fingertip with the hanging fingernail. Before the proverbial iodine and bandaging, my father decided that a hanging fingernail would not heal properly, so he did what needed to be done. He borrowed a pair of pliers from my uncle's toolbox and ripped off the nail. And so I held sweet Fido, cowering, shaking, and nipping at me a few times, and ultimately got all the spines out. He recovered fully.

Fido was a mutt with long black and white fur. In the evening when we had a fire, he would curl up on the warm hearth and sleep. We burned dry white pine, a very poor firewood. Instead of a slow, hot burn, white pine flares up quickly, and resin causes it to pop and send out hot embers. Every once in a while, the room would be filled with the smell of burning hair as an ember would burn its way through Fido's coat, luckily never reaching his skin, and he remained snoozing obliviously.

Constantly worried about fires, Miss Patton made sure we always put the screen over the fireplace when we left it. I liked to use the third-floor bedroom, but she would let me stay there only if I could demonstrate that I could escape in the event of a fire. I showed her how I could lower myself out the window, drop onto the roof over the porch, and jump down to the ground. Eventually, her fears were realized. Long after we'd stopped going there, the house burned to the ground. Wendy has since visited and shared with me a photo of the small replacement cottage. It's a sad little building on a site where a grand old home full of memories once stood.

Early School Days

Living things fascinated me from an early age, but as I grew older, my interests broadened. On entering my teenage years, I thought of myself as a lover of math and science generally. The first external recognition of that came in grade seven when I won the prize for "High Achievement in Arithmetic Scholarship Examination." The problem I solved was the calculation of the surface area of the end of a barn, ostensibly to determine how much paint it would take to cover it. And the prize? A copy of *Nicholas Nickleby*[9] by Charles Dickens. That may seem like a strange math prize, but I had told my teacher I was reading *The Pickwick Papers*,[10] a book that simply appeared in our living room one day.

My father had "lifted" it from a house he worked on. I don't know why he did that. I imagine he liked the distinctive English nature of it. As far as I know, he never read it. But I did. I used to come home from school in the afternoon and lie down on the sofa in the living room. It was a bulky, soft sofa—grass green brocade with a sheen. I would stretch out full-length with a cushion under my head, oriented with my feet away from the window so the light would fall on the pages. The afternoon sun came pouring down through vertical Venetian blinds, casting light and dark bars across everything in the room. One day, I read the following line:

> He chuckled, roared, half suffocated himself by laughing large pieces of beef into his windpipe, roared again, persisted in eating at the same time, got red in the face and black in the forehead, coughed, cried, got better, went off again laughing inwardly, got worse, choked, had his back thumped, stamped about, frightened his wife, and at last recovered in a state of the last exhaustion and with the water streaming from his eyes.

That's when I laughed out loud for the first time while reading a book, and it was Dickens's *Nicholas Nickleby* that prompted it. It wasn't all great, classic literature that I read. I'd go to the local public library and check out Hardy Boy mysteries and just eat those up. I was the only one in the household of six doing any reading other than my dad, who read *The Montreal Star* as he waited for my mother to serve supper at 5 p.m.

When I graduated from elementary school and won that math prize, my teacher advised me to register for the Latin program at Rosemount High School. Students were placed in one of three tracks there: Latin, Science, or Mechanical. Not realizing that these were academic levels and liking science so much, I registered for the Science stream in spite of my teacher's advice. I spent grade eight in class with tough guys who smoked, hung out with girls, and had no interest in schoolwork, whereas I was a "good boy" and actually liked school. The typing teacher taught my ninth-grade biology class, and I knew more about biology than she did. The staff instituted a new system when I entered tenth grade, Subject Promotion, and I was able to take the more challenging classes with the former Latin students. The teachers were concerned about my ability to master Geometry as a tenth grader, but I had a good spatial sense and got the highest grade in that class. It was instructed by Miss DiPierro, who had actually taught my mum, and they shared a warmth for children. She was a kind old lady who would put her hand on our shoulders as she moved between the desks, looking down at our work. In my early teaching days, I felt quite parental towards my students and similarly found myself placing my hand on their shoulders as I circulated the lab. I stopped that innocent behavior as our culture became sensitized about sexual harassment.

In grade eleven, I entered a science fair with the encouragement of my physics teacher. I don't recall how I chose my project, but I ended up making a thermopile, a device that converted heat into electricity. Modifying a plan in the March 1963 issue of *Popular Mechanics*,[11] I twisted together the ends of fifty pairs of wire strands, one iron and one constantan (a copper–nickel alloy), and my dad found someone to weld the entwined ends. I arrayed the wires in a radiating circular pattern on a fourteen-inch disk of asbestos-impregnated cement board, with the welded junctions in the center and the outside ends twisted to join adjacent pairs. It looked like a daisy and formed a continuous circuit except for two free ends, one iron and one constantan. When I heated the center with an alcohol burner, the heat gradient between the hot center and the cool outer edge of the disc generated a tiny electric potential in each wire pair, and with fifty pairs attached in series, the setup provided enough voltage to run a tiny electric motor and turn a little fan. I built something, it worked, and that felt good. The project demonstrated a scientific concept but did not test a hypothesis. It did do something, though, which was better than my girlfriend's project, a model of a nuclear reactor. That was the state of science education back then.

I did not win anything, but an engineer from an electronics company encouraged me and sent me some semiconductor material. I didn't do anything with it, and I had no idea how he expected me to use it. The other encouragement I got was from my teacher; he asked me to demonstrate my project to the class. I was always a shy kid and got very nervous in front of groups of people. For the demonstration, I used a much hotter Bunsen burner, instead of the usual alcohol burner, as my heat source. At one point, I reached over to adjust my apparatus and passed my bare arm right through the transparent flame. It didn't hurt, but the intense smell of burning hair filled the air, the same smell that had filled the room when an ember fell on Fido's coat. I continued with shaking hands, hoping nobody noticed. If they did, no one mentioned it.

Undergraduate Studies

I applied to one university, McGill. I had been on campus twice before. The first time, my father took me to the Redpath Museum. I was four at the time, and a large gorilla in a glass case filled the entryway, standing upright with its right arm uplifted and holding onto a tree branch. It looked so threatening and I cringed, not wanting to go in for fear it would break out through the glass and kill us. Dad finally convinced me it was not alive, and in we went.

The second time I was on campus was as a senior in high school. I visited as part of a program for prospective students. Jack Berrill, an eminent marine biologist, warmly welcomed me to his office. I recall being enthralled by a large circular glass tank of seawater with gelatinous walnut-shaped bodies circulating in the current. He saw my fascination and informed me that they were comb jellies (ctenophores).

Being accepted to McGill was my Mum's dream come true. When I arrived as a freshman, I knew I wanted to study science, but physics and biology attracted me equally. Feeling lost in class, I quickly understood I was not cut out for physics. But I really shone in biology and felt at home. I may have done better in physics if I had had better professors, but I did well in biology, no matter their quality.

My real forte was lab work. I grew up watching my father clean fish and butcher deer. He also did all the maintenance around the house, so I learned to use tools and be good with my hands. Because Dad worked in construction, speed was important, so his work was rough. In contrast, I was very fastidious. I worked slowly but carefully, a trait I shared with my Ukrainian grandfather and one that translated beautifully into the lab. In freshman year, the teaching assistant commented on the precision of my fetal pig dissection and drawings. For the final lab test in Vertebrate Zoology in sophomore year, we walked into the lab and were faced with rows of trays, one at each station stretching down the long black benches. Each tray held a small preserved shark head. The assignment was to dissect out the ten cranial nerves and produce a labeled drawing of them. I lowered my head and went to work, focusing intently while inhaling formaldehyde fumes. I exposed every nerve precisely. I made the edges of the cartilaginous cranium, where the nerves emerged, perfectly straight. It was a piece of art! I raised my head at the end of three hours of work with a great sense of satisfaction and a precise, properly labeled drawing.

Being good with my hands did not necessarily make me a good scientist. I did not do very well in Organic Chemistry, but then sixty percent of the class failed. The course was well-known as the cut-off point for pre-med students. I did get an excellent grade in the lab component of the course, though. I could follow directions precisely and make the apparatus exactly as described in the lab book, so I got precise, clean results. Did I know what was going on? No. It was straight cookbook. I guess I could have been a technician in any science lab, but only in biology did I really have a thorough understanding. I feel so fortunate to have ended up in a field I love and am good at.

A Close Call

Being at college was more than academics. As a commuter student, I missed out on much of the social life, but not all. In addition to the occasional dances, I was an active member of the Outing Club, and I took part in various sports: badminton, squash, basketball, and the varsity track team. But one intramural sport almost got me in serious trouble—floor hockey. A uniquely Canadian version of this game, it was played on the hardwood floor in the gym. The goals were similar to ice hockey nets, the sticks were basically straight broom handles, and the puck was a thick ring of sewn felt layers. With five players and a goaltender, it was much like ice hockey. The puck was handled by placing the end of the stick in the central hole and pressing down as you moved about, looking for an opening to shoot at the net or pass to a teammate. Light checking was allowed.

Some of my friends and I put together a team to play in the intramural league. We won enough times to be in the playoffs but were eliminated in the first game, the only one I remember. It was early evening, and we assembled on the court in the gym. We were a rag-tag bunch of skinny boys, and across the court stood our opponents, a team from the dentistry school. They were solid, muscular men, somewhat intimidating. When the game started, it was clear those guys were playing a physical, hard-hitting game. We were being pushed around and not getting anywhere, but for a while, there was no score. At one point I saw an opening and ran to my left and took a shot at goal. Somehow, all the pieces came together perfectly: the puck flew towards the net and slipped over the goalie's shoulder just under the crossbar. My team cheered and said it was a Bobby Hull shot. Bobby Hull was the hottest scorer in the National Hockey League at the time.

And with that shot, I was a marked man. The next time I had the puck, I was illegally cross-checked. A player ran at me with his stick held horizontally in two hands and delivered a heavy blow to my chest, knocking me down. I completely lost control of myself. I had what you can call a "fit of rage." I ran at him and swung my stick as hard as I could at his head. To my everlasting thankfulness, I missed. Had I connected, I might have cracked his skull. And then what? I could have gone to prison, convicted of assault. My academic career would have been over. The question would no longer be biology or physics.

Choosing a Path

I gave up physics after my freshman year to focus on biology, but what kind of biology? I had no idea. I was in a big university, and I had no advisor. I took a variety of courses in a vague attempt to get a rounded background in the subject. As I entered my junior year in the fall of 1965, I enrolled in Peter Grant's course called "Ecology." It took very little time for me to realize I had arrived. I had found the subject that truly fascinated me. Ecology was an obscure branch of biology back then. Karen's father, in his speech at our wedding reception in 1969, said: "When we heard that Ray was studying ecology, we all ran to our dictionaries."

Peter Grant, known to me as Dr. Grant at the time, was a very energetic young English ecologist early in his career. He later became famous for his groundbreaking work on the evolution of finches on the Galapagos Islands.[12] Recognizing something in me, he gave me the opportunity to participate in fieldwork between my junior and senior years. He designed an experiment, assembled the materials, helped me set it up, and left me to carry out the project. What was the project? A field experiment to determine if spiders reduced the population sizes of their prey.

I spent that summer at the Gault Nature Reserve on Mt. St. Hilaire, a small mountain rising out of the flat agricultural St. Lawrence River Valley. It was covered by a beautiful forest consisting primarily of large, mature sugar maple and beech trees. My study took place in the thick leaf litter on the forest floor. Fallen leaves provide a source of nutrition for decomposing bacteria and fungi that support a complex food web of invertebrates (pill bugs, millipedes, springtails, ants, mites, pseudoscorpions, centipedes, and many others) in which spiders are the top predators. We marked off four twelve-foot squares on the forest floor. Two of these we surrounded with sheet metal fences partly buried in the soil, and two were unfenced. One fenced plot was the experimental, and the other three were controls. On the experimental plot, I removed all the spiders and censused the remaining invertebrates for three months.

To remove spiders and census other invertebrates, I picked up handfuls of leaf litter and placed them in two nested white plastic wash basins. I had cut the bottom out of the top one and replaced it with a screen. I vigorously shook this setup, causing the animals and tiny litter pieces to fall through and collect in the bottom basin. I then sorted this material, identifying and counting all the creatures and removing the spiders in the experimental plot. By carefully marking off the plot with string, I moved through the area systematically, processing all the litter.

Each day that I censused a plot, I made myself a lunch and hiked up to the study site. I filled the basin, shook it, returned the litter to its original location, and sat on the ground with my back against a tree to work through the animals and fine litter pieces. Repeating this over and over all day long, I processed the entire plot. It was lonely work. I had a small portable radio and listened to tinny music for company. When the task was completed, I walked back to the dorm, ate dinner, and hung out for a while with a couple of graduate students also residing there. One was studying the microclimate within the forest, and the other was studying the small mammal community. In the evening, we sat around, played darts, and talked. Once, we heard some rattling outside from a fifty-five-gallon drum that served as our trash can. A couple of raccoons were foraging in there. To see if we could catch them, we snuck up and slammed the lid on the can. OK. So now what should we do? We didn't know, so we lifted the lid, and two fat raccoons shot out of there and waddled off as fast as I have ever seen raccoons move. Most evenings were less eventful, and I sleepily turned in quite early. What did I see when I closed my eyes? Spiders. Not scary spiders. I handled spiders all day long with no problem. I was comfortable with them. I just saw so many throughout the day that they were imprinted on my brain.

During the set-up period, Peter Grant walked to the study site with me a few times. I learned to avoid following too closely behind him as we bushwhacked through the trees because the young beech branches were like steel springs. A couple of times a bent branch hit me in the face as it whipped back to its natural position. It really stung, but there was no real damage. However, on one occasion, as I followed him down a steep slope, he stepped on a stone and felt it move. He turned to warn me, but it was too late for me to stop. The stone rolled out, my foot slipped, and my ankle slid along a sharp edge, creating a deep gash. Blood poured out, but there was nothing to do but keep walking. When we got back to the parking area, he drove me, and the blood-soaked sock on my foot, thirty miles to the emergency room in Montreal. The doctor repaired the cut with a dozen stitches, and I still carry the scar from that episode.

I did not appreciate it at that point, but the spider project was part of a new wave of research in ecology. Until then, most of the field research was observational, sometimes quantitative, but not manipulative. Researchers were just beginning to actually perform controlled experiments in the field, and Peter Grant was actively involved. At the time, ecologists were debating the effect of predators on their prey populations. Some believed that predators took "excess" individuals and thus did not have an effect on prey populations, while others believed that predators truly controlled the populations of their prey. In our experiment, the density of prey in the plot without spiders was significantly higher than in the controls, thus providing evidence that predators have a subtractive effect on their prey. As a graduate student at Yale, I gave a seminar about this study in a class taught by G.E. Hutchinson, probably the greatest ecologist ever. (He introduced so many new concepts that initiated large areas of research). "Hutch," as we called him, but not in his presence, underlined the significance of the spider work. I was so flattered to unexpectedly receive such a remark from an intellectual giant we all worshiped.

Peter Grant and I worked together to write the paper. The project was eventually published in *Ecology*,[13] the world's most prestigious ecology journal, with me as the first author, and I was off on a research career. As I look back at that paper now, I see so many weaknesses in it and know that it would not pass peer review today. The field has progressed so far in terms of methodological precision and statistical sophistication. But it was significant in its time, and I have Peter Grant to thank for setting me on the path to a lifetime of gratifying research.

I never told Peter how much he influenced me. I wish I had. On a couple of occasions, a past student revealed how much a course she took with me had meant to her. Each time I heard that, I was surprised and gratified. Students don't let on. Teaching is rewarding and would be even more so if we received more positive feedback.

In my senior year at McGill, I applied to five graduate schools in Canada. I had no intention of leaving the country. I was accepted in all but my first choice, The University of British Columbia. I decided to attend The University of Saskatchewan because Dr. Grant told me I would be working with the foremost ecologist in Canada, Richard Miller. After my admission, I received a letter from Dr. Miller with the news that he had accepted a position at Yale University, but he had me set up with another faculty member at The University of Saskatchewan. Alternatively, he said, I could apply to Yale to work with him. I did that, was accepted, and prepared to move to New Haven. It was 1967, a turbulent time when anti-Vietnam War and the civil rights movements were peaking. Many cities, including New Haven, were experiencing violent demonstrations. My parents asked why I wanted to go so far away. Montreal to New Haven is 370 miles, whereas Montreal to Saskatoon is 2,120 miles. I guess they meant so far culturally.

I met Karen during my freshman year at McGill. You won't believe this, but it was at a hockey game. She was sitting next to me and dropped her glove. I bent down to retrieve it, presented it to her, and we started a conversation. It's true! At first, I was more interested in her friend, Wendy, and dated her a few times. We did not click, though, and I started dating Karen. She was my girlfriend throughout college. When I went to New Haven, she went to a teachers' college in Toronto, and we parted, getting together during vacations in Montreal. When she visited me in New Haven, we talked about the future, decided to marry, and phoned our parents with the news. Looking back at photos taken at the time, I see we were so young, twenty-three, still naïve, immature, unsophisticated. I, a budding biologist, and she, a budding math teacher. Two linear thinkers.

Encountering Reefs

Don a mask and snorkel, put your head in the water, and swim around seeing the submerged world clearly while still breathing—what an appealing concept! I had starry-eyed images of snorkeling since I was nine or ten but had to wait until I was a graduate student when Karen and I, along with three others, took a trip to the Florida Keys during winter break. My dream finally came true, and I was not disappointed.

A fellow graduate student, Jim Sharp, organized the trip. His parents regularly wintered in their trailer at a campground and motel on Big Pine Key, and we stayed for a week. It was a wonderful time. We snorkeled around the bridge pilings and marveled at the fluorescently colored juvenile reef fish darting about. They seemed to have no momentum as they stopped, started, and turned instantly, jerkily. I had never before experienced clear tropical water both from below while snorkeling and above from a boat. On Christmas day, I recall planing along the surface in a power boat over transparent water, zipping over bulky brown sponges that seemed shallow enough to hit but weren't. Jim's dad had traps set out for stone crabs, and we harvested these, or, to be more specific, their claws. Jim showed me how to twist off the one large claw of each crab and release the crab to grow another. We felt that we were engaged in a sustainable endeavor, but I have since learned that it takes at least three years for a crab to regrow a claw fully, and since large crabs are near the end of their lifespans, few would live to develop another harvestable claw. In my ignorance, I enjoyed the sweetest meat I have ever tasted.

Jim was an entrepreneur, so he arranged for us to collect live reef fish for delivery to a local pet shop in New Haven. We had the use of his father's boat and regularly visited the reef at Looe Key, five miles offshore from Big Pine Key. Looe Key was not an island or key. It was just a submerged shallow reef in the Florida Reef Tract, but it was unusually beautiful, and many boats could be found anchored there, each with a brood of snorkelers surrounding it. The waves were bigger than anywhere else we had snorkeled, and I swallowed seawater a few times. There's nothing like salt water in your stomach as you sit in a bouncing boat to make you feel queasy. At one point, a reef shark swam by with its dorsal fin projecting from the water. Everyone got into their boats, as did I. If it happened now, I would stay in the water and watch the sharks swim by.

It was the snorkeling on Looe Key that changed my research direction. I was working with toads in New Haven and continued that project to complete my dissertation, but my future work would be on coral reefs. Two specific observations at Looe Key fascinated me. The first was the abundance of fish and other animals, including coral. I saw a system predominated by consumers with very few plants to support them, just a small amount of algae. This contrasted with terrestrial systems where the animals are embedded in the plant-dominated environment, and it raised a big question in my mind: Where is the food that supports all this animal life? I have since learned that many reef species feed on plankton imported from the surrounding open ocean, a high level of photosynthesis occurs in the form of single-celled algae embedded in the coral tissues, and the high turnover of the ubiquitous short filamentous strands of algae means that they produce a lot of food, even if there is little there at any point in time.

The second observation, beyond the abundance and diversity of colorful fish, was the fact that they seemed to be going about their business without concern for my presence, even a few feet away. On land, if you want to watch bird behavior, you have to build a blind and move about stealthily to avoid alarming them. On the reef, the fish were feeding, defending their territories, and guarding their nests right in front of me. This conduct was not as big a mystery, but it suggested they would be excellent subjects for studies in behavioral ecology. Had I thought more about this lack of fear, it would have been difficult to explain knowing what I did then. I now understand that the answer lies in the tremendous density of fish on the reef. If little fishes responded defensively to every larger fish that swam by, they would always be hiding, unable to go about the behaviors necessary for survival and reproduction. They are very perceptive and react to their predators while ignoring the more abundant herbivores and clumsy snorkelers.

We collected lots of fish using the quick-acting anesthetic quinaldine and hand nets. We packed them in plastic bags half-filled with seawater and topped off with oxygen from a bulky green cylinder. These were stuffed into coolers, and we were off for the trip home. We left on Friday evening and drove straight through, arriving at New Haven mid-morning on Sunday. We delivered the fish to the pet shop and kept some for ourselves in a large tank I had built using the plate glass salvaged from a broken tabletop in the Divinity School dining hall where we sometimes ate. This was in the early days of silicone cement when all-glass aquaria were a novelty. Up until then, aquaria, like my father's, had steel frames and slate bottoms for strength. My aquarium was not the all-glass tank that you see today; it had a slate bottom, a carry-over from the old style. The fish thrived for a week or so but then declined and died, one by one. At the time, we did not understand the nutrient dynamics of salt-water aquaria. A buildup of ammonia waste had become toxic.

Floyd Connor and I shared an office in grad school. We also shared an advisor, Rick Miller, who was himself searching for a new research project. The two of us wrote a long proposal trying to convince him of the value of coral reefs for research. He ultimately went in another direction, but after completing my dissertation on toads, I moved on to studying coral reefs and even did a field study with Floyd. The rest of my research career was spent investigating the behavioral ecology of reef fishes.

Learning to Scuba Dive
(In Spite of Myself)

When I was still living on Laval Street, I would occasionally go "swimming" in the local indoor public pool. *Bain* means bath in French, and in the bilingual tradition of Quebec, the words "Bain" and "Bath" sat over the entrance of the building's imposing stone front. I don't know if it was ignorance or some form of arrogance, but we spoke of Bain as the name of the place, so we would say: "Are you going to Bain Bath today?" In any case, I went there frequently by myself, trying to learn how to swim. A neighbor told me he learned to swim when an older friend threw him in the deep end. I preferred a less frightening, slower approach. In increments, I eventually got to the point of doing an awkward doggie paddle, but I at least got there. I never learned to swim efficiently, though. I can do a respectable-looking crawl, but something about my technique is off. I do not move through the water nearly as fast as others, but I take that as an advantage in the sense that I get a better workout than others when swimming. And to this day, I never learned to dive.

At fourteen years of age, I stood there in my swim trunks at the edge of the swimming pool. My toes gripped the tiled rim as I leaned forward with outstretched arms, preparing to dive, preparing to dive, preparing to dive—and nothing. I couldn't allow myself to fall headfirst into the water.

There are times when your head tells you one thing, and your gut says the opposite. The unthinking gut reaction is an evolved mechanism that protects us from dangerous situations. Survival may depend on an instantaneous reaction, not a time-consuming, methodically worked-out response. For example: "See snake; get out of there immediately." We know the fear of falling is built in because experiments by psychologists Eleanor Gibson and Richard Walk in the 1950s with naïve human infants, kittens, and the young of several other animals showed that they would not crawl or walk off a visual cliff.[14] The visual cliff was an apparent but not a real precipice, and the babies could venture over the edge without falling because the drop was under a clear acrylic sheet. Nevertheless, the various babies would stop when they came to the apparent edge. No experience necessary.

The interplay between instinct and calculated response varies considerably in people. We know that some folk are daredevils. For example, in 1974, Evel Knievel jumped the Snake River Canyon in a homemade rocket. Over the years, he had 433 broken bones in over 20 crashes.[15] Others, like me, are overly cautious. We have recently discovered that there's a biological basis for this difference.[16] I never learned to dive into a swimming pool because I couldn't bring myself to fall headfirst. However, I did manage to overcome a different built-in response. After Floyd and I decided that coral reef fishes would be our next research focus, we took a scuba diving course in 1969.

The course took place in the Athletic Center at Yale and was taught by Peter Dingwall, a technician in the Physics Department. We met twice a week over a semester. Peter's goal was to scare the bejeezus out of us. He wanted us to be fully aware of the dangers of scuba diving so that we took all safety precautions seriously. For example, he showed us slides of cadavers that had been run over by power boats. I can still see the rows of slices the rotating propellers made through the bodies.

But Peter backed up his fear-mongering with precise instruction and exercises to help us develop the skills and self-confidence to handle emergencies without freaking. He started by giving us swim tests. The tests involved laps in the pool and breath-holding, underwater swimming. Several students did not pass and were dismissed from the course. I did fine with these tests but came close to failing the last, which involved treading water with your arms above your head for three minutes. The requirement was that your elbows had to stay above the surface. Karen, being a woman, had enough body fat to float easily and passed the test without effort. I, on the other hand, was quite thin then, so my greater density meant I would sink. As a result, I had to kick powerfully to keep my elbows out and would tire before the end. I solved the problem by keeping my arms above the water but submersing my head to get a little added buoyancy. I lifted my head occasionally for a breath, then plunged it back under. I met the requirements technically, so I passed.

The pool exercises were designed to make us feel comfortable in the water under unusual circumstances. One exercise was to swim the length of a pool, breathing through a snorkel while not wearing a dive mask. People and other mammals have a basic reflex to hold their breaths when their faces are wet. However, in an emergency while scuba diving, if the mask is knocked off or floods, the adaptive response is to continue breathing if you still have the mouthpiece in place. This exercise was designed to prevent panic. When I first tried it, my breathing instantly shut down; I could not make myself inhale. Trying again, I forced myself to sip some air and rapidly became comfortable, and it became a total non-issue. While diving, I now readily flood my mask or even take it off to clear the fog that sometimes develops on the glass lenses.

It was all quite macho at the time, but my clearly not-macho wife Karen took the course, too. She was a better swimmer than I, very comfortable in the water, and easily mastered all the exercises. There were no pressure gauges then, so we could not monitor the amount of air we had in our tanks. Instead, when breathing got difficult because of low air pressure, we would reach behind and pull on a ring at the end of a long rod attached to a valve, and that would free up some extra air to get us to the surface. The valve was controlled by a lever, which we confirmed was in the "up" position before each dive. If that lever was inadvertently pushed to the "down" position during the dive, which could easily happen if you bumped into something, you would pull on the ring, and nothing would happen. You were out of air—time for buddy breathing.

To initiate buddy breathing, you would have to swim over to your partner and make the out-of-air signal, which was drawing a hand across your throat. She would remove her mouthpiece and hand it over to you. You would take two breaths and hand the mouthpiece back to her. She would take two breaths and hand the mouthpiece back to you. This would continue as you both slowly swam in tandem to the surface. We had to practice buddy breathing so we could carry out the procedure smoothly and safely if needed. All very simple—no panicking, very controlled—in theory. But would you give up your only source of air to a panicked partner, trusting that he would relinquish it when it was your turn to breathe? It worked just fine when we practiced it in the pool, and fortunately, we never had to use the technique in an emergency. Nowadays, scuba equipment comes with two regulator mouthpieces, so if your partner needs air, you can simply pass them your secondary regulator (the octopus), and they can safely breathe from your tank.

As with swimming, the greatest test was the last one. I had to place my fins, tank, and regulator on the bottom of the pool. The air valve was turned off. As I stood on the side of the pool above my equipment, the instructor handed me my mask. The inside had been covered with aluminum foil, so I could not see when I placed it over my eyes. The test required me to dive down, find my mouthpiece, turn on the air valve, and start breathing while I put the tank on my back, then hunt for my fins and put them on. To make the test even more challenging, the assistants were allowed three "harassments" while I tried to get organized at the bottom of the pool. For example, my air was shut off, and I had to reach behind my head to turn it back on. When I approached my fins, they were moved away, and I had to feel around in a search pattern until I found them. Eventually, I got myself organized and properly fitted, then rose to the surface to remove my mask and see again. I passed.

Diving certification requires that the students have two open-water dives. Our course was in the spring semester, so our open-water dives occurred in mid-May. We loaded ourselves into a van and drove to Rhode Island. On a windy, brisk, sunny day, we got into a boat and headed away from shore. On anchoring, it became clear that Peter had miscalculated the tide—a strong current was flowing. Nevertheless, we donned our black wetsuits, struggled with our tanks, and checked our regulators a hundred times. Then, it was over the side and into the chilly, chilly water. Because of the current, we had to descend hand-over-hand along the anchor line. I looked down and saw a row of timid new divers, their bodies waving like flags as they held on with both hands for dear life. Then I became one of them. The current was much reduced at the bottom, and we were able to explore a little while we stressed over the mechanics of diving. It was the most difficult dive in my whole career.

That was then. Today, you can go to a tourist resort, get an hour of instruction, and go below. The equipment is much safer and easier to use, but I question how such an untrained person would respond if something went wrong.

The Brain and the Gut

I have frequently said that a dive course is a great maturing experience. I was forced to overcome my built-in phobias. It made me a stronger, more confident person. The instinctive fear of falling is still strong, though, and a much more difficult challenge for me occurred in the fall of 2015 when I was visiting a rainforest preserve in Peru. The associated eco-lodge had erected a zip line from treetop to treetop high in the canopy, and I really wanted to see the forest from that perspective. To get to the starting point, I ascended a giant emergent tree without trouble. After all, I started on the ground and was hoisted up incrementally while in a harness and connected to a steel cable. The problem occurred when I had to jump into the void and zip between trees.

I sat on the edge of the platform, harnessed and connected to the sturdy steel cable by a carabineer and, for extra safety, an independent tandem zip line by a second carabineer. Through the leaves, I could see the forest floor one hundred feet below. I knew that all was secure and entirely safe. But it took probably five minutes before I could force myself to make the leap. Five long minutes during which the guides were losing patience and other tourists were tolerantly waiting their turn. My brain said: *Go already. It's no big deal. It's safe. Go!* My gut countered with: *Don't kill yourself.* So, I remained frozen until I could get my brain to overcome my gut—not eliminate the gut feeling, mind you, but jump out in spite of it. My heart raced like never before. After I finally vaulted, I zipped along with slightly less trepidation, but I never relaxed until I reached the platform in the next tree.

Dan on zipline in Peru. Photo by author.

That experience, the experience so many seek, was totally wasted on me.

Zip lining was not the only time I needlessly felt I was going to die. On a family vacation trip to Florida, I went on a roller coaster with my son Justin, who was ten years old at the time. The "Scorpion" had a track that formed a vertical loop so that riders reached fifty miles per hour, and the centrifugal force kept them in place as the car careened through the top upside down. Once on board, I had no decision to make; the car was moving, I was in it, and there was no way out. I had no problem sitting there relaxed and facing the sky as the car slowly cranked upwards with that characteristic clanking sound of roller coasters. But the vertical drop scared the bejeesus out of me. My limbs froze, my heart raced, my brain turned to jelly, and I felt I was plunging to my death. After a couple of turns, we came rattling into the exit station, and I stepped out with shaky, rubbery legs only to hear my son exclaim: "Let's do it again, Dad." Needless to say, I did not entertain his request. To this day, forty-year-old Justin remembers that nobody would go on the roller coaster with him.

Similar conflicts in choice also occur in biological contexts. Walking in a park in Toronto, I came across a bee fly. I had never seen one before, but I knew about them. These flies get their names from their fuzzy black and yellow abdomens that cause them to resemble bumblebees. I wanted to capture the specimen to illustrate the phenomenon of mimicry to my students, but I had no collecting equipment. My only option was to catch it with my bare hand, which can be done if you move slowly, very slowly, until you are really close and make a quick strike. I did the slow part but had tremendous difficulty bringing myself to close my hand on it. My head said it was perfectly safe. After all, I had seen it lightly flit from flower to flower like a fly. In contrast, bees drone on, flying more slowly and seeming to be heavier, having more momentum. But my gut said: *Don't be stupid. You'll get a painful sting.* My head prevailed, and I caught the fly, but it was not an easy thing to do.

I'll never dive into a swimming pool. I'll never get on a roller coaster again. And while I have not tried, I doubt I would ever bungee jump or parachute out of a plane. Others readily do these things. I know I have a very logical mind and understand they are reasonably safe activities. Clearly, the power of my inhibitions is greater than those of many other people.

Although the gut reaction was an adaptive response in bygone days, it is not necessarily so in our modern world. We are programmed for a past environment, what evolutionary psychologists call the environment of evolutionary adaptedness. The wide range of responses in people today represents the variation that provides the raw material for evolution. To the extent that the variation is due to genetic variability, there is a potential for evolutionary change. Perhaps in generations to come, these responses will disappear, but I can't see us turning into pure logical beings.

Back on Solid Ground

In my early forties, I trained to run in the New York Marathon. For a year, I ran miles and miles, including a few twenty-milers, all over Ossining and surrounding towns. I was slim, fit, and feeling good. During a short run on a sunny October afternoon, I saw an elderly man slowly jogging along the road ahead of me. He looked like he was struggling as I effortlessly caught up to him. I slowed down as I passed him because I didn't want to zip by and make him feel bad about his condition.

I was born to run. As a child, I ran everywhere. Well into my thirties, the only way I would climb stairs was quickly, two steps at a time. I wasn't aware of my talent until the eighth grade when our gym teacher took us out to the track on a spring day and timed us running a mile. I left the whole class behind and ran it in five minutes and twenty-five seconds. I played on the soccer and basketball teams in high school but was unexceptional in those sports. Every spring, however, I was a leading member of the track and cross-country teams, so it was natural for me to join the track team when I went to McGill University. We had daily training sessions, and I did well. I joined a track club to continue training throughout the summer. I had achieved an identity. I was a known runner in the area. When I visited a store to purchase a new set of track shoes on McGill University's account, the young man fitting me recognized me and said that I had beaten him in several races. In my sophomore year, I won the silver medal for the indoor 600m at the Canadian championships in Winnipeg.

At the Quebec track championships, I anchored the mile relay team. In this event, each member runs one lap, a quarter mile, and passes the baton to the next. I was the anchor, the last runner. As I waited in the passing zone, watching my teammate approach ahead of the pack, my legs felt heavy, without energy. This was normal; I always had this feeling at the start of races. I was always a slow starter and a strong finisher. I took the baton with a large lead and ran through the first turn, striding easily at a good pace. Down the backstretch, I became aware of the crowd cheering louder and louder. A runner had caught up and was abreast of me—I took off. He had spent much energy closing the distance between us, and my cruising for the first half of the race left me with a lot of reserves. I powered through the second turn and down the straightaway, crossing the finish line with a greater lead than I had started with. The crowd cheered, and we celebrated, having broken the provincial record for that event.

There were lows among the highs, of course. At the international track meet in Toronto, each event had several tiers. My coach entered me in the elite half mile. I don't know what his thinking was, but I found myself at the starting line amid world-class runners. I was nervous, not my confident self, even though I had the preferred inside position. Did I say I always got off to a slow start? Not this time. At the gun, I shot out like a rabbit and led the field through the first lap and a half. Down the backstretch, I began to tighten up and was passed, left behind by several runners. As I labored past the finish line, I looked back to see there was still one runner behind me. I had not come last. But with my head turned, and in my fatigue, I tripped and fell onto the cinder track. The next day, I read an article in the *Toronto Globe and Mail* describing the race. A line from that piece is still imprinted in my brain: ". . . and then Clarke died."

I continued running road races through middle age, mostly ten and twenty kilometers. I did well but was never near the top. I was fit, fit to the point that I just had to run. One Friday night, dark and pouring rain outside, I felt I needed to move, so out I went. I also played tennis and later squash as I got older. But my back was sore and the twisting of tennis caused me pain. My energy level dropped. I no longer took stairs two at a time. Now, I have arthritis in my left knee. And even less energy. I often take a nap after lunch. I sometimes have trouble remembering words. My concentration is not what it used to be. I'm thankful for pickleball, a game that can be played by the physically compromised. But I feel sore and tired for a few hours after playing six or seven games. I don't run anymore.

Recently, at the age of seventy-six, I went for a walk in Central Park in New York on an early November afternoon. It was after the schools let out, and several groups of young men were running along the trails. I assumed they were athletic teams on training runs. As they passed me, I noticed that each was led by a few easy runners, smoothly striding and chatting away, followed by a tight group obviously working and not talking but maintaining the pace. Strung out behind, a few stragglers with legs and arms flailing labored to keep up. The antelopes at the front of these groups reminded me of my college days when I was at my peak. I thought about my present condition and felt the loss, but not in a big way. It is what it is. I can't change it. And anyway, I'm in pretty good shape for my age. I recalled that old man so many years earlier I'd slowed down to pass. He was probably no older than I am now, thinking about himself I as do. I didn't have to slow down for him. It is what it is.

The Nature of Science

Have you ever watched puppies chasing each other, tumbling over each other, rolling on the ground, biting each other, but not for real? They're playing, of course. Unlike other behaviors that have a survival or reproductive purpose—foraging, courting, nest building, and predator avoidance—play does not have an immediate purpose. It's just fun. It's an end in itself. Of course, if you take a longer view, play is practice for all the skills that young animals will need to be successful adults.

Like other mammals, we humans play as children. But we are different in that we continue to play as adults. We persist in doing things that are ends in themselves. Why else do we play video games, crossword puzzles, card games, and sports? The makers of these products generate money, but the consumers derive some form of enjoyment, nothing more. Play becomes valueless as animals mature, and few adult mammals play, but we humans persist. We can hypothesize that this pattern has value in fostering problem-solving, a major adaptive trait of our species. So, we grow up and keep playing. The games change, and we spend less of our time doing it, but we never stop.

We also play with ideas. For example, in the late 1980s, as I was working thirty feet deep on the fringing coral reef on the north side of St. Croix, an object grabbed my attention. It was midday, and in the crystal water under the bright tropical sun, the reef looked like a movie set lit by klieg lights. Before me, sharp as a tack, on a blade of elkhorn coral, was a long, mottled brown mass, the width of a pencil and a little shorter. It was gently rocking back and forth with the swell. I thought that this might be a marine worm mimicking fish feces to avoid being eaten by a fish. This thought came to me because I knew that some caterpillars resemble bird feces to avoid being eaten by birds. It was another example of how my training in terrestrial ecology gave me insights into my research in marine biology. So, building on my success in extrapolating from land to sea, I set out to test the idea by waving my hand at the long, mottled brown mass. The swirl of water washed it off the coral, and it completely disintegrated. It was fish feces.

Caterpillar of the Giant Swallowtail, Papilio cresphontes. Photo by A. Reago and C. McClarren, CC, https://creativecommons.org/licenses/by/2.0/legalcode.

Here's the point: science is a form of play; it's sophisticated, but nevertheless, play. It shares with tumbling puppies the fact that it is an end in itself. That's why I do science. It's fun. It's problem-solving. I engage in curiosity-driven science, sometimes called pure science. It is extremely satisfying and an end in itself. Other people do what's called goal-oriented science or applied science. They are solving defined practical problems, like cures for diseases or alternate forms of energy production. But they, too, derive great satisfaction in the discovery, aside from the practical value of their work.

For some, science is simply meaningless unless there is a clear economic benefit. A memory sticks with me from 1974 when I was working at the Lerner Marine Lab on Bimini in the Bahamas. When the wind was too strong for us to dive on our field site, we would sit on the dock in the lagoon and map the movement patterns of tiny bright yellow and blue damselfish called beaugregories. We could easily observe them from above as they defended their territories. One such day, my colleague Floyd and I were sitting on the edge of the dock with our maps on our laps, recording the movements, when a family of tourists came by. The father asked what we were doing, and when we explained the project, he turned to his family and said: "So that's how our tax dollars are being spent." We didn't engage him, and he walked off.

Pure science provides value to society, too, but in a less immediate way than applied science. Think about the scientists working in the 1940s and 1950s trying to understand the molecular nature of inheritance. They were motivated to understand how cells function. They were curious. They simply wanted to know. Their work culminated in the famous photo, spread around the world, of triumphant Watson and Crick with their model of the DNA molecule. Today, molecular biology is the underpinning of our medicine and many industrial processes. Who would have known this in 1954? Like play, curiosity-driven science provides value in the long run.

More Play

Sometimes, scientists like to play outside the lab, too. It was Saturday night, 1968, and the graduate students at the Yale School of Forestry and Environmental Studies were having their annual Halloween party. The room was crowded with costumed partiers drinking punch and talking animatedly when in walked a slender student dressed in black pants and a black tee shirt. Moreover, his shiny dark hair was plastered down, and black streaks ran down his face. That student was me—I had oiled my hair and used black eyeliner to draw gobs running down my face. Most people dress as pirates, Romans, celebrities, and the like. But here was me, doing my strange thing. What was I? An oil slick.

I don't know why I did such an unconventional thing. It probably has something to do with my desire to make a statement about costume parties. I am critical of them for reasons that are not clear to me. At some level, I am saying: "You want a costume? Let me show you a costume." I truly am perplexed by this attitude. In any case, I have never gone to a party in a conventional costume. I've always appeared dressed in outlandish ways that confound others and require my explaining what I am supposed to be. Depending on who's receiving that explanation, I get a reaffirming nod or a puzzled look.

In 1979, during the first few weeks of my yearlong sabbatical stay on St. Croix, I arrived at the lab Halloween party in shorts, a tee shirt, and flip-flops, the standard tropical lab attire. But my head was crowned with a small light atop a tiny cardboard tube. I had rigged a flashlight bulb with some wires that ran into my pocket, where they were connected to a battery. What was I? Buck Island. Buck Island sat a couple of miles off the north coast, where the West Indies Lab was located. It had a navigational light sitting on a tower that flashed in the night sky and was visible from the lab—a constant presence. Denny Hubbard, the faculty member hosting the party, told me later that he would have honored me with the worst costume award, but I was new then, and he did not know me well enough.

The following fall, I was invited to a department Halloween party at Cornell University. I wore a crown consisting of a thin piece of sheet metal with about twenty small nails sticking up from it. When asked to explain my costume, I answered: "I'm a spinyhead blenny." That got knowing looks from nobody. So, I spent the evening repeatedly explaining that a spinyhead blenny was one of the fish I studied. It lives in tight-fitting holes in coral, and its spine-covered head can plug the hole.

The last time I went to a costume party was several years later in New York City. In preparation, I crumpled several sheets of newspaper into balls and glued them onto an old baseball cap. After spray-painting the whole thing green, I had my costume. I was a head of broccoli, of course. For some reason, all my costumes were centered on my head. I never dressed up further.

I am a generally quiet person, not good at making cocktail party conversation, and usually trying to avoid drawing attention to myself—except at Halloween parties. Maybe those costumes were lures. Just as an anglerfish cannot chase down its prey but uses a lure to entice them in, maybe I adapted to my hesitation in starting conversations by creating lures to draw people in. "What are you supposed to be?" is a conversation starter—and I got others to initiate it. Although the costumes were icebreakers, I did not consciously design them for that purpose, as far as I can recall, but who knows what motivations lurk in the unconscious mind?

Where did the inspiration for my getups come from? I have no idea. You can ask the same about how scientific hypotheses enter our heads. Consider the benzene molecule. In 1862, the German scientist, August Kekulé, knew the number of carbon and hydrogen atoms in the molecule but could not arrange them into a chain. While dozing in front of a fire in Ghent, Belgium, he had a dream of a snake seizing its own tail and suddenly saw that the benzene molecule was a ring.[17] That was one of the first times anyone had hypothesized a ring structure for a molecule.

Fieldwork

Some people are attracted to fieldwork by a love of Nature and a desire to be close to living things. Others have a purely intellectual interest in theory, and fieldwork is simply data collection to test theories. For me, it is both. Fieldwork allows me to be a physical organism in touch with my animal nature. Data analysis allows me to be an intellectual being engaged in abstract thought. It's very close to the Athenian ideal of developing a balance of mind and body. More basically, it's another form of bringing home the bacon; the collection of data is not unlike a hunter-gatherer foraging for food.

Terrestrial field biologists immerse themselves in the environment they are studying, but they are moving in the usual manner of bipedal primates. Among marine biologists, blue water scientists study the open ocean from ships, which are little pieces of land moving through the sea. But many coastal marine biologists quite literally immerse themselves in the environment they are studying. Humans have been free diving for more than a thousand years. Most impressively, the Bajau of southeast Asia can remain submerged for over five minutes and reach depths beyond 200 feet.[18] A recent innovation, scuba-diving gear has been around for only 70 years and was not widely used until the 1960s, when technological advances made it less cumbersome. When I learned to scuba dive in 1969, all the equipment was black, and we carried big knives strapped to our legs—very macho. Today, you can outfit yourself in pink or turquoise or probably even chartreuse if that's your style, but you still need to be fit enough to manhandle a 35-pound air tank plus lead diving weights. Fortunately, once below the surface, the experience of effortlessly moving through a three-dimensional environment is closer to flying than walking.

In my work with tiny fishes living in cavities in the coral, I spent most of my time on the bottom with my head near the reef surface. My diving partners would sometimes ask if I had seen that shark or that large parrotfish or those dolphins, and my answer was inevitably "no." Once, a remora even tried to attach to my back without my knowledge. If I wanted to get an "aerial" view of the study site, I simply had to keep a little more air in my lungs, and I would slowly rise like a balloon. The viscosity of water and the need to conserve air means that divers move slowly, and the experience is very relaxing. It's even possible that our metabolic rates slow down as well. The diving reflex of seals seems to be a modification of a general mammalian property in which, among other things, heart rate is slowed. Humans have this, too, and in the Bajau, it's more pronounced. In spite of this calming feeling, however, there was always work to be done.

The point of fieldwork, of course, is data collection. The value of the data is largely determined by the care with which it is gathered. Because of the need for statistical validity, a lot of data must be collected and this can make fieldwork repetitive, thus boring, even in beautiful locations. However, it is the currency of our field and the collection of a good data set is in itself satisfying. The great value that scientists place on data was humorously exemplified by U.C. Santa Barbara behavioral ecologist Bob Warner while at the West Indies Lab. I was approaching the dock one afternoon as he was leaving, my Boston Whaler full of wet gear scattered around and his full of dry gear neatly arranged, plus two graduate students. As our boats passed, he was standing tall in the bow and challenged me with the following: "Your data or your life."

Sometimes You Have to Get Creative

Before immersing myself completely in reef studies, I needed to complete my dissertation research on toads. I had decided on toads after hearing my professor of evolution recounting the conditions in which different species would hybridize. In 1970, while searching for a study site, I visited an isolated piece of abandoned land in Hamden, Connecticut, where truck tire tracks had filled with rainwater and toads had laid their eggs. The little mini-ponds were teaming with tiny coal-black tadpoles. They formed dense masses in the shallowest sections, so shallow that their shiny wet backs projected through the surface tension, each ringed by a bright reflection of the sky in the curved meniscus. I suspected their packed black bodies would absorb sunlight and heat up, so I measured the water temperature

Toad tadpoles in tire tracks. Photo by author.

within these masses and found it to be significantly warmer than in unoccupied sections of equal depth. The warmer water was important because it would raise their metabolic rates and result in faster growth, an adaptive feature because they had to develop quickly and metamorphose before the puddles dried up. As far as I knew, nobody had documented this before.

I returned to the site one evening to see how the tadpoles distributed themselves when the sun was not shining. While making my observations, three young men came walking by. They asked me what I was doing, and when I explained it to them, they asked: "What good is that?" Their tone was quite provocative, so I snottily answered: "You wouldn't understand." That was not my finest moment, and I'm thankful I didn't pay for my arrogance with a bloody nose. If now I met the young man I was then, I might not like him much. He was rigid and insecure in some settings, condescending in others. But, I'm sure I would have enjoyed talking biology with him.

Fowler's toad. Photo by author.

The primary site for my toad study was the Yale Golf Course. The data collection occurred at night during closed hours, and I had a key to get me past the main gate. The basic procedure involved walking around with a gas lantern and searching for toads in a systematic manner. By walking the identical route in the same manner, I had a standardized sample and could compare the number of toads, their locations, and their activity at different times. One August night in 1971, as I was peacefully strolling in the cool of the evening, serenaded by the scratchy calls of katydids in the trees, a set of headlights and a roaring engine sped in my direction on the road. The car veered onto the grass and headed straight towards me, suddenly braking five feet away. Both doors simultaneously flew open, two cops jumped out, and they demanded I stop. At least they didn't have their guns drawn. I went through the explanation of my presence there at night, showing them my key, and satisfied, they drove off. My halo of light once again enveloped me in the night as my racing heart slowly calmed down, and I resumed my work.

Being a good scientist involves a lot more than having the humility to acknowledge the tentative nature of our understanding. It also involves creativity in inventing ways to gather data. For example, to fully understand the population biology of an animal, it's important to determine the size of its home range, which is the area in which the animal moves during its regular daily activities. The only way to do this in an unbiased way is to tag individuals with some kind of transmitter that allows a researcher, using an appropriate receiver, to locate the individual at any time, wherever it might be. Sufficiently small radio transmitters were not available in 1970 when I was studying toads, and even today's models include an antenna, which is not suitable for a soft-skinned burrowing amphibian. The only alternative was to use a radioactive isotope that could be inserted under the skin.

By reading the published literature, I found that the isotope most frequently used for this purpose was tantalum-182, which has a half-life of 114 days and emits powerful gamma rays that can be detected from a workable distance. I submitted my research plan to the Health Physics Committee at Yale University. In that plan, I demonstrated that I knew enough about the movements of my subjects to be quite certain not to lose them and the isotopes they were carrying. The Committee did not approve my use of tantalum-182 because if I did lose a toad, the material would be in the environment for too long. However, they approved of my plan with the proviso that I find an isotope with a much shorter half-life that would result in its quickly decaying and becoming harmless should I not be able to recover it. After a careful search of the data on isotopes of a variety of metals, I decided that manganese-52 was the only appropriate material. It has a half-life of 5.6 days. The Committee approved.

Now, all I had to do was obtain some manganese-52. That isotope was not available commercially, but it could be produced by bombarding chromium-52 with positrons. The Atomic Structure Lab at Yale had a linear accelerator that could do that. All I had to do was provide them with chromium targets, and they would do the calculations, generate the right energy level of positrons for the right time period, and voila, I'd have my manganese-52. I needed to obtain some very pure chromium to avoid producing unwanted isotopes during bombardment. I imagined I could buy some chromium wire and cut it to 5 mm lengths. However, it turns out that chromium is a very hard and brittle material. It cannot be formed into wire, and the only way to get it in very pure form is to have it deposited from a solution onto an electrode until a thick layer is built up. Fortunately, such chromium was available commercially. When my chromium arrived, it looked very much like peanut brittle—broken pieces about a quarter inch thick, smooth on one side and covered in little bumps on the other. A talented technician in the Engineering Department formed an electrode with a 1mm hole in it, placed a chromium chunk into an oil bath to remove the generated heat, and slowly lowered the electrode onto the chromium. As the electrode moved through the chromium, the material was vaporized, leaving a 1mm diameter rod.

Plated pure chromium (left) and sealed manganese-52 rod (right). Scale in millimeters. Photo by author.

I delivered ten of these rods to the Atomic Structure Lab, where they were mounted at the end of the accelerator and magically zapped. On arrival to pick up my newly produced manganese-52 rods, I entered a cavernous space littered with complex machinery. Cables were running everywhere. The technician in charge prepared to remove the material, but several short-lived secondary isotopes had been produced, so the area was "hot," meaning quite radioactive. A primary way to minimize exposure to radiation is to reduce the time you are close to the source. The technician exhibited tremendous agility in zipping up the ladder to the hot target area, unlatching the cover, removing the material, and quickly descending. I walked away with my precious isotope in a lead container. Dense materials absorb radiation best, hence my lead container and the lead apron you wear when getting dental X-rays.

Back in my lab, I removed each tiny rod from the container and slid it into a glass capillary tube. By then, the short-lived isotopes had decayed, and I worked speedily but without other protection. With a Bunsen burner, I fused the ends of the capillary tube, so now I had my manganese-52 sealed in glass, which prevented any possibility of some dissolving and escaping into the body fluids of the toad. That evening, using a large bore needle and wire plunger, I inserted the material under the skin of 10 toads I'd established had well-defined home ranges. The plan was to locate each toad once an hour for a week. I set up a small tent in a wooded part of the Yale Golf Course so that I would be on site the whole time. Day and night, I took catnaps until the alarm went off, and I mapped the locations of the toads.

One piece of information I had provided to get the approval of the Health Physics Committee was that the toads were strictly nocturnal and would not come in contact with the patrons of the golf course. But one afternoon, as I set out to relocate a toad that had sheltered under an old railroad tie used as a step leading to a raised green, I was horrified to see two golfers sitting there having a chat, their testicles only a foot away from the toad and its radioactive tag. I was about to approach them with a warning, a most awkward situation, when they stood up and walked off, saving me from a tough conversation and them from an increased chance of producing mutated sperm.

The tagged toads confirmed my capture-recapture method of determining their home range and also provided me with a sample of their daytime hiding places, something I could not learn in any other way. I recovered all the tags and returned them for proper disposal. All of that took place in 1971 when radionuclides were much more freely used than today. Although I was a lowly graduate student, being at a big university provided access to tremendous technology. That was another time, both in terms of abundant resources and the acceptance of isotopes as benign tools. I cannot imagine getting permission to release radioactive toads onto a golf course now.

Sometimes You Get Hurt

Fieldwork can be very uncomfortable. When I first worked in the Caribbean in the mid-1970s, the long-spined sea urchin, *Diadema*, had not yet experienced its die-off and was extremely abundant. With a three-inch shell and copious, black six-inch spines radiating in all directions, they resembled spherical porcupines. Those spines were covered with living tissue that released venom when broken. The urchins occurred in densities up to 100 per square yard on many coral reefs. In the early 1980s, the species experienced massive mortality across the whole Caribbean and have only partially recovered since.

My work with blennies meant that I spent all my time close to the reef surface. Hardly a dive went by without my bumping into a long-spined urchin and feeling an intense burning pain. I have to confess I lost control at times, turning in anger and smashing the offending but innocent creature. This crack-up inevitably led to a mad rush of little fishes called wrasses that greedily fed on the exposed flesh. The pain in my punctured skin usually subsided in a quarter-hour. The pieces could not be removed the way a wood splinter could because they were constructed of calcium carbonate and crumbled when manipulated with a needle. Often, a piece of spine embedded in my skin left a black tattoo-like mark for a week or so before being absorbed by my body.

Back on the reef, rock-boring sea urchins provided an even more challenging situation. These urchins have reddish-brown spines about an inch long, not as sharp as those of the long-spined sea urchin but pointy enough and considerably stouter. Each rock-boring urchin excavates a shallow cavity in the soft carbonate rock and shelters there during the day. At night, they emerge onto the rock surfaces and graze the algae that grow there.

Lamb Bay, on the east end of St. Croix, has a beautiful white sandy beach, the kind you associate with tropical paradises. It also has a white sandy bottom with coral patches that supported the diverse fish communities I was studying. Between the sandy beach and the sandy bay bottom is a strip of beach rock, a ten-foot wide irregular rock band that has to be crossed to get into the deeper water. This band's bumpy, pock-marked surface is usually covered by about six inches of water. It harbors a dense population of rock-boring urchins, perhaps fifty per square yard.

When we did our fieldwork in 1979, Karen and I would enter the water at about 2 p.m. At that time of day, the urchins were nestled in their cavities, and walking over the beach rock was no problem. The issue occurred on exiting. After 4:30 p.m., when the sun was lower in the sky, the urchins began to emerge from their cavities. We stood up, knee-deep on the sandy bottom, seawater dripping off our wetsuits and down our legs, weighted down by heavy scuba tanks on our backs, lead weight belts around our waists, fins looped over our wrists by their straps, and assorted recording equipment, nets, and other paraphernalia cradled in our arms. We wore neoprene booties that kept our feet warm but provided little physical protection from spines.

We remained there for a minute or two, staring at the beach rock studded with emerging urchins, building the courage to make the crossing. There was no way we could get from the water to the beach without stepping on urchins and getting bits of broken spines embedded in our heels. You can't step lightly, prance, or tip-toe when you are so heavily burdened as we were. It was a heavy plod, and worse, the rock surface was very irregular and the footing difficult, so we would occasionally lose our balance and have to stomp down to regain it. When we would step on an urchin, the spines penetrated our heels and broke. "Damn it." We tried to pick our spot for each step but ended up getting caught, and it hurt.

A month after returning from St. Croix, my body exhibited a new strange pattern. Every night around bedtime, I experienced an intense itching around my anus. What could that be? The first step in finding out was to look at the irritated site. I used a mirror and saw several tiny white worms crawling around. I had been infected with pinworms. These are nematode worms (commonly called roundworms), and this species infects a person when its eggs are ingested, usually from contaminated food. They live in the gut, and when mature, the females migrate to the anus while the host is sleeping, deposit their eggs, and die. It's the eggs that cause the itching. When the host scratches the site, the eggs are transferred to the fingernails, where they can be ingested to continue the infection, or they can be transferred during food preparation and infect another person. This is one of many examples of parasites dictating the behavior of their hosts to insure successful reproduction. Strict hygiene, especially frequent hand washing, interrupts this cycle in about one month. And, fortunately, it did.

Sea urchin spines and pinworms are just an annoyance. I once faced a truly life-threatening experience. It may have started when I scraped my shin on a brain coral in Belize or during the subsequent work in Port Aransas, Texas. That part of the project involved plankton collection on the lab's pier on the ship channel. In my spare time at the end of the day, I sometimes walked out on the channel's jetty to see what kinds of fish the many people were catching there. I don't know in which of these places it happened, but I caught something, too: the insidious *Vibrio vulnificus* bacterium, a denizen of warm oceans.

It was no big thing at first—I had a slight red swelling on my left knee. I felt a little off but kept working in the lab. The next day at lunch, I felt terrible and weak. I lay in bed and did not go in that afternoon. Carly, my student assistant, carried on with the experiment, wondering where I was. When Ed got the word, he asked his wife to help. Nancy was a nurse at the local hospital. She immediately drove me to the ferry, and we crossed to the medical center where I was diagnosed with cellulitis, which could develop into the frightening necrotizing fasciitis, commonly called flesh-eating bacteria. The doctor prescribed an antibiotic, and Nancy drove me back to Port Aransas.

I was scheduled to return to New York in a few days, but Nancy told me to cancel my flight and stay in Port Aransas for further treatment.

Expecting better treatment in the New York area and not fully appreciating the severity of my condition, I ignored her advice and got on the plane. I spent the whole flight bent over in my seat, shivering violently. It was agony. On landing, I managed to debark and shuffle to the baggage carousels, where Brigid was waiting for me. She took one look at me, grabbed my bags, and drove me straight to the hospital emergency room. I was immediately admitted.

The next day, a doctor put me on an intravenous antibiotic drip and used a felt marker to outline the swelling so he could monitor its progression. Two days later, with the red swelling extending beyond the magic marker, he stood at the foot of my bed with two other physicians. They looked at me and were rubbing their chins. Oh, that was scary. Finally, the doctor said that the Vibrio bacterium was resistant to all the antibiotics they'd tried. He followed that news with: "We'll try the big gun." At that time, there was one antibiotic that still worked on bacteria that were resistant to all the others. In a couple of more days, my condition improved, and I was released after spending a week in the hospital.

When I had an infection in 1953, I was given penicillin. The bacteria at the time were naïve and responded immediately. By 2004, various antibiotics were so overused that many bacterial strains had evolved immunity to them. Fortunately for me, my Vibrio responded to the only last-resort antibiotic available. And here I am, still kicking.

Boating for Science

Over the years in Bimini, St. Croix, and Belize we always used the same boats for fieldwork—either thirteen- or sixteen-foot-long Boston Whaler skiffs, ideally suited as research tools. These boats were wide and had flat bottoms, making them very stable, and their low sides made it relatively easy for us to pull ourselves into them from the water. They had simple boards as seats. We headed out to the reef with a pile of equipment covering the bottom—scuba tanks, flotation vests, weight belts, regulators, fins, masks, snorkels, nets, measuring tapes, meter sticks, cameras, writing slates, buckets, and net bags filled with smaller tools, such as syringes and calipers. Sometimes, there was little room for our legs. After diving, that pile became a mess; everything was wet, and we sometimes had live specimens to protect in the jostling jumble. Most trips were uneventful, but I made perhaps a thousand of them over the years, so inevitably, some interesting things happened.

At the beginning of my reef research career, Karen and I were novice scuba divers, and I was a novice boat operator. We did not have the confidence that comes with experience. I knew what to do but had to think explicitly about each step and felt vulnerable out on the open ocean in a small boat. On one of our deeper early dives in Bimini, I released the anchor and watched it drop to the bottom. The water was crystal clear, and I was struck by how small the anchor looked when it finally hit the sand sixty feet below. At that point, a six-foot lemon shark cruised past the anchor. We considered aborting the dive since we knew lemon sharks were known to attack humans. But the shark was headed in a straight line and moved off, out of sight. After some thought, I decided it was safe. We descended and did our work uneventfully. I saw sharks rarely during my years of work and never felt threatened. Rather, I was always struck by their slow, sinuous undulations as they cruised so efficiently over the reef with a smoothness unmatched by the frenetic swimming style of most other fishes.

In contrast, our other big-fish encounter occurred in three feet of water. We were swimming over white sand along the edge of some rocky outcrops to census the fishes living in that zone. Suddenly, out of nowhere, right in front of my face, appeared a gigantic flank. It filled the water from the sand to the surface – a wall of gigantic, shiny-silver overlapping scales. I immediately knew it was a tarpon, no threat at all. But when it had passed, and I looked for Karen to acknowledge our good fortune, I found her hiding behind me.

The most dangerous incident on a boat occurred during my first field season of reef research. Motor maintenance left something to be desired at the Lerner Marine Lab on Bimini in the summer of 1973. The outboard motor on the boat assigned to me was unbalanced, so if I let go of the tiller at cruising speed, the motor would rapidly slam to the left. I had to hold the tiller firmly to maintain a straight course. Here's where my inexperience showed. Returning from our field site,

Karen and I happily headed home in the late afternoon when a small, long-handled net, sitting in a bucket with the mesh end up, caught the wind. As it headed over the side, I made a grab for it with my right hand, the hand that until then had been on the tiller. The motor instantly shot over to the left, and the boat lurched hard right. Enter Newton's First Law of Motion, which states in part that a moving object will remain moving in a straight line unless acted upon by an external force. Karen, peacefully sitting on the center seat, continued in a straight line when the boat made its sudden sharp turn, hurling her overboard, where she hit the water with a giant splash. I was thrown down but remained in the boat, circling around and around. Karen was a good swimmer and remained calm, but I panicked as I wrestled the motor to get control of the boat without running over her. I finally lowered the throttle and was then able to slowly maneuver beside her and help her into the boat. I don't remember what happened to the net.

One bright morning during the following field season, while motoring out to our study site with Floyd, something came flying out of the water, hit him in the chest, and fell to the bottom of the boat. We immediately stopped, searched the boat, and found an eight-inch reef squid flapping about. Its tentacles were flailing, the fins at the other end of its body were beating, and its color was rapidly pulsing between psychedelic states. I was familiar with the behavior of flying fish, which, when chased by predators, build up speed and glide above the water to reenter far from their pursuers. It appeared the reef squid had a similar behavior in this instance, reacting to the fast-moving boat as a predator. We put the squid in a bucket of seawater and took it back to the lab to photograph for teaching purposes before releasing it.

Unlike the flying squid, seeing frigatebirds fly was no surprise. Frigatebirds roughly resemble seagulls, but instead of sharing the white coloring of most seabirds, frigatebirds are black, and they're big with long, narrow wings that span almost eight feet. The wings are held in a bent position when the birds are soaring and, combined with their long, deeply forked tails, create a sinister appearance. They remind me of the Nazgul in J.R.R. Tolkien's *Lord of the Rings*.[19]

Male magnificent frigatebird. Photo by author.

Usually, I see these birds flying at some height, but on one occasion, while I was motoring alone to a study site on St. Croix, I saw a frigatebird flying low with something spherical and white in its beak. I was watching the bird because of its unusual behavior when it dropped the object. Curious, I headed over to see what it was. There, floating high on the water's surface, was a three-inch puffer fish. It was inflated with air, a tight ball of projecting spines. In a couple of minutes, it managed to expel the air and swim for the bottom, its fins beating furiously like little propellers. Usually, puffer fish inflate with water to prevent predatory fish from swallowing them. Frigatebirds are not divers, so I speculated that this fish must have been close enough to the surface to be picked. Perhaps it was caught by a diving bird and then stolen. Frigatebirds are known to harass other birds, forcing them to drop their prey, which the frigatebird then picks out of the air. In any case, by inflating with air, this little puffer avoided being swallowed and lived to spend another day on the reef.

When taking out a boat at the West Indies Lab on St. Croix, we always filled out a form indicating the time, destination, and purpose of the trip. This was routine; I did it hundreds of times. However, one Saturday night, there was a party at the lab, and a few of us, having downed several beers, decided to take a trip to Buck Island. I, being a visiting faculty member, had the authority to sign out a boat, and at eleven o'clock, five of us headed into the darkness, through a narrow gap in the reef and across two miles of calm ocean to the Island. We hung out there for a while; some walked along the beach while the rest of us lay on the sand, watching the stars and listening to the waves lapping on the shore. After an hour or so, we motored back across the open water toward the lab dock. The narrow gap in the reef was harder to find in this direction because the lights on shore reflected on the water's surface, making it difficult to read the waves and find the marker buoy. Although I was really drunk, this all happened without incident; I was very familiar with the local waters by that point. However, the next morning, the maintenance supervisor made a snide comment to me about taking a boat out at 11 p.m. for R & R. What could I say? I felt suitably chastised and never did anything like that again. But unlike my first boating adventure, I had pulled it off skillfully.

Sometimes You Get to Live a Dream

Gliding over white sand without scuba gear in deep blue light fifty feet below the surface, I was relaxed and remarkably unencumbered—complete freedom—a sensation similar to what I had experienced as a child when I used to dream I could flap my arms and glide above the streets of Montreal. But this wasn't a dream. I was swimming between Hydrolab, an underwater habitat I'd been living in for six days in 1978, and the tank rack. The living quarters were too small and its hatch too narrow to put on scuba gear inside, so we'd take a breath, enter the water, and swim about fifteen feet to a rack, which held full scuba tanks placed there by the surface crew. Upon reaching the rack, I chose a tank, opened the valve, placed the regulator in my mouth, and took a breath.

The time any diver can spend underwater is limited not only by the amount of compressed air stored in the scuba tank, but by the accumulation of nitrogen in their blood. Going up is the dangerous part. That's when divers can get the bends, caused by nitrogen bubbles forming in the blood as the pressure is reduced. As a diver, I have to start surfacing from a hundred-foot dive after only twenty-five minutes, which is not nearly enough time to do all the things a researcher might want to do, such as setting up equipment for an experiment or observing fish spawning behaviors that continue for some time. But in Hydrolab, I could live deep and dive long. By living at fifty feet below the surface, I could spend unlimited time at one hundred feet below the sea's surface because the bends do not occur between these depths.

Hydrolab, operated by the National Oceanographic and Atmospheric Administration, was a steel cylinder, eight feet in diameter and sixteen feet long, designed specifically for saturation diving. It sat on the

Hydrolab and Its Umbilical Cord. Photo by Rod Catanach

white sand plain of Salt River Canyon on the north coast of St. Croix, halfway between two coral-covered walls a quarter-mile apart. The sand plain begins in about twenty feet of water and plunges to over 4,000 feet. To prevent the sense of being trapped inside an empty gasoline tank, Hydrolab had a circular four-foot acrylic viewing port at one end of the habitat that flooded the inside with blue light during the day. We entered the chamber on the underside, squeezing through a two-foot diameter hatch leading into "the trunk." Compressed air provided a dry living environment under the sea. It was pumped in from a floating barge above and bubbled out of the open hatch, keeping the water out and maintaining the internal pressure equal to that of the surrounding water. In addition to air, the umbilical cord connecting the chamber to the surface barge also carried fresh water, electricity, and a cable for radio communication. Our lives depended on that filament and its connection to the machinery on the barge bouncing on the waves above.

The trunk was a five-foot vertical steel cylinder with a second hatch at the top. Only three and a half feet in diameter, it functioned as an airlock for exiting after decompression. It also served as a wet porch. It was there that we removed our wet suits, took a cold-water shower, and dried off before reentering the living quarters. We were told to notify the surface crew if the shower was comfortably warm—not a good thing, as it meant the air compressor was overheating. Our wet suits stayed in the trunk and never fully dried out. Throughout my stays in Hydrolab, I never got used to putting on cold, clammy neoprene.

During storms, passing waves created oscillating pressure that caused the water surface in the trunk to rise and fall. At these times, the sound of water rushing in and out, like in a bellows, filled the inside of the habitat, amplified and sharpened by the steel walls. Our eardrums were alternately stretched and relaxed by the oscillating pressure waves—hearing muffled, hearing clear, hearing muffled, hearing clear. I was amazed that the different pressures of the peaks and troughs of the waves did not even out at that depth.

Despite being left alone on the bottom of the ocean, we never felt vulnerable or in a dangerous situation. We were free to do our work without worry because we had a support crew at the land base: a doctor who had worked as a physician for navy pilots, an engineer who had worked on the deep-sea submersible Alvin, and several very experienced divers. Before submerging, we spent several days in training and testing. We were given physical exams and physical work, such as hefting bulky eighty-pound twin scuba tanks into boats in the baking sun and taking a distance swim test. We were taught about navigating underwater and how to respond in an emergency. During that period, the competence and confidence of everyone impressed me. I felt that no emergency would overwhelm our team. They knew their stuff.

The trunk in Hydrolab. Photo by author

Author in trunk of Hydrolab, 1981. Photo by Jim Morin.

On each of our missions, a crew of four "Aquanauts" occupied Hydrolab, but there were only three bunks. One person was supposed to be awake at all times in case of emergency. The person on each two-hour watch radioed the shore base once an hour to give the "all okay below." Exhausted by diving for eight hours daily, it was hard to stay awake while your three colleagues were sleeping. Back then, waterproof Casio digital watches were new, and most scientists found them to be effective and cheap alternatives to the bulky, costly traditional analog dive watches. One member of our crew solved the sleep problem by setting the alarm and taping the compact watch to his head so he could nap in a chair and still make the call on time. The interior of Hydrolab was furnished with a stainless-steel counter running down one side and bunk beds on the other, separated by a narrow aisle covered with green fake grass carpet. Pipes and cables ran everywhere, and there was a constant hiss of compressed air. We had a mini fridge, a hotplate, and lots of freeze-dried meals. It was all very tight. We ate on our laps while sitting on the lower bunk. Nobody could move unless another made way. After several days I really missed fresh food and radioed for lettuce and tomatoes. I received a laugh at the other end. Fresh lettuce was hard to come by on St. Croix at the time, but a crew member somehow found some and "potted" it down. The "pots" were the pressurized vessels designed for spray painting. They kept the contents dry and had valves to equalize the internal and external pressure; otherwise, we would never be able to remove the lids on pots sent down. In this manner, we received and sent out dry things. It's also how our clothes, books, and computers were delivered when we moved in.

Inside Hydrolab. Jim Morin in forefround, Myra Shulman in middleand Ed Brothers in background. Photo by author.

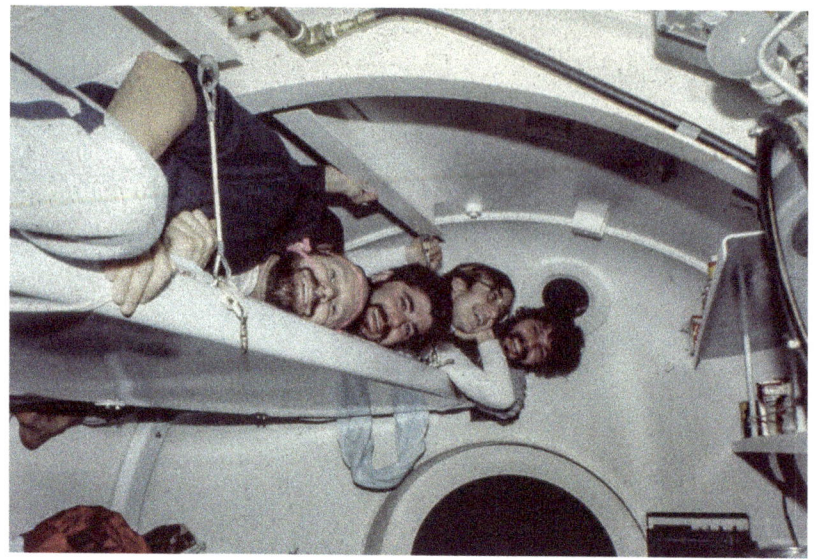

Silliness in top bunk of Hydrolab. Front to back: Jim Morin, Ed Brothers, Myra Shulman, and author. Photo by Jim Morin.

The bunks were stacked against the wall, one above the other, and could be swung up out of the way to provide a sitting surface during the day. They were lowered at night for sleeping, but the upper bunk barely came down because of the curved top of the chamber. Whoever got that bunk would have to squeeze in and take a comfortable position because once he or she slipped in, there wasn't enough space to turn over. For amusement during a rest period, we crammed all four people into that bunk, heads, legs and arms dangling over the edge, for a photo. We also amused ourselves in other ways during down times. Ed Brothers, a renowned specialist in using fish ear stones to determine their age, was an avid angler who fished at every opportunity. Using a cube of cheddar cheese, he baited a safety pin attached to some twine and dropped it through the hatch. We weren't paying much attention to him until he yelled, "Got one!" and stood there with a yellowtail snapper wriggling on the end of the line. He immediately released it. This silliness may have been more than a need for entertainment during rest periods. There is a well-known phenomenon called rapture of the depths (nitrogen narcosis) caused by excessive nitrogen dissolved in the blood of deep divers breathing compressed air. Those experiencing it have a sense of extreme well-being and lose judgment; they often make bad decisions, like offering their regulators to fish. We had some colleagues visit by diving from the surface to work with us saturated divers. One friend commented to me afterward that I was swimming "sooo slowly." Now that I think of it, I might have spent the whole week in a low state of nitrogen narcosis.

Because of the unlimited bottom times, we used double scuba tanks, which provided up to ninety minutes of dive time, but even that was not enough for some research procedures. The surface crew stashed extra tanks near study sites so we could change out without returning to the habitat a couple of hundred yards away. At the site we used most often, the tank drop was in the midst of a colony of very territorial three-spot damselfishes. Each time I settled down to change tanks, they started biting me, and I flinched with each nip. They didn't cause any harm, but nevertheless, instead of being a slow, methodical operation, tank changes were always rushed as I tried to get away from those fearsome fish. It was very humiliating to be chased away by a four-inch fish, though I took some comfort when I learned other colleagues had done the same.

Spending so much time diving took its toll on us beyond extreme fatigue. My mouth became quite raw from the rubbing of the regulator for so many hours. I also developed serious heartburn. Hours in the horizontal position, sometimes even working in a head-down position, allowed the acidic orange juice I'd had at breakfast every day to seep back into my esophagus. The project doctor told me that the valve between the stomach and esophagus is not as tight as the one at the other end of the stomach. Until then, I had no idea that I depended on gravity to keep things down. As a result of the experience, I lost my interest in orange juice for years afterward.

Our sole contact with the outside world was by shortwave radio. For us aquanauts, that consisted of the radio person at the land base for our operations. Our conversations were generally all about the state of the habitat and planning dives and deliveries. However, when George Dale was down below with me, it was the first time he'd spent away from his two small children. He arranged with the people at the land base to patch him into the telephone system, and he made a call home. He instructed his three-year-old daughter to say "over" when she finished speaking. The rest of us could hear the conversation over the loudspeaker and were touched when she said: "I love you, Daddy, ova, ova."

There were no toilet facilities at Hydrolab. Relieving oneself required entering the water, breathing through a long tube (a hookah), and getting privacy by placing oneself out of view of the ports. Apparently, there is some nutrition in human excrement because the fish quickly learned to assemble and wait for the goodies to emerge—even the elegant French angelfish, a slow-moving jet-black dinner plate with yellow-edged scales. In spite of the groupies, this was an especially serene moment, floating with my trunks down and connected by that long air hose, away from my ever-present companions and the hissing noise of the habitat. It served as a moment for reflection, to think about the project at hand and the fact that I was actually living fifty feet below the surface. Although life in Hydrolab had become routine, looking back at the habitat, connected by that thread, gave me a feeling of amazement that it was all happening.

Research in Hydrolab

In my early thirties, with a full brown beard, I saturated for a week in Hydrolab on three occasions over four years, twice for projects I was awarded in the competitive granting process and once as a technician on a colleague's project. I shared the first project with George Dale (full brown beard), a biologist from Fordham University. Each time he greeted me, he would flick his eyebrows, a built-in non-verbal gesture from his deep evolutionary past. We studied the role of light in determining the remarkably synchronized changing of the guard at dawn as nocturnal fishes enter shelters in the reef and diurnal fishes emerge from those same shelters. The reverse happens at dusk. We were attempting to show that specific fish consistently use the same shelters and timed their comings and goings with great precision. Some individuals varied their timing by as little as three minutes over six days.

The second time in Hydrolab, I was helping John Ebersole and Les Kaufman. Deaf in one ear, the result of a brain tumor in his early twenties, John (full brown beard) had shifted from studying birds to reef fish because you don't need hearing to study fish. Les (full brown beard) was a very perceptive observer who made many innovative contributions to marine biology. The goal of that project was to contrast the differences in reef fish community composition on the north and south walls of Salt River Canyon. The north wall consisted of abundant corals forming overlapping plates projecting from the vertical canyon face, whereas the south wall consisted of hemispherical corals scattered on the sloping algae-covered rocky substrate. We hoped to address an ongoing controversy about the importance of physical structure in determining the distribution of species.

I shared the final mission with Myra Shulman (no beard), a fun-loving biologist with a knack for working hard without seeming to. Our project was a study of the feeding behavior of garden eels in relation to water currents. Garden eels are sixteen-inch fish that feed on passing plankton. Their tails remain embedded in the sand while their bodies project straight up as they search the water for food. They live in large groups, and from a distance, a colony of garden eels looks like a bed of seagrass. Moving slowly and fluidly, they calmly went about their business, but the closer I approached them, the deeper they sank into the sand until they disappeared. As I sat at the edge of a colony, I would see no eels nearby. As my gaze moved further and further away, they would appear as little nubbins, and then finger-like projections, and pencils, and finally fourteen-inch eels, heads facing into the current, dancing back and forth as they captured their passing prey. We were particularly keen to discover the importance of water currents in delivering the eels' food. Were the large expanses of white coral sand uniformly suitable for them, or did variations in water currents dictate the locations of the colonies?

All of these projects were connected to ideas floating around in the 1970s and 1980s about coral reef fish community structure. There is still a controversy about the importance of randomness as compared to precise, orderly response to the environment. Are reef fish avoiding competition by precise specialization in the spaces they occupy, or are they broadly overlapping in their use of space and avoiding competition in other ways? All my immediate colleagues and I fell on the order side of the debate. Of course, as in all other either-or controversies, we now understand that reality is a combination of the two hypotheses.

Joys and Sometimes Dangers of Hydrolab

Perhaps my most spectacular experience while below was on the mission in which we were studying dawn and dusk reef fish changeover. After completing our observations one evening, George Dale and I started the long swim in the dark back to the habitat. We were following the rope laid out to guide us home as a thunderstorm was pelting the surface fifty feet above, and we could clearly hear the shushing sound of the raindrops hitting the water. We also saw lightning, silent lightning, so we turned off our flashlights. To our amazement, the agitation of our fins caused luminescent algae to light up. So, here were two divers slowly swimming in the dark, each with a ten-foot trail of light behind him. Periodically, a lightning flash flooded the whole reef in daylight, but it instantly returned to dark except for the two glowing trails.

Not only did I have the most aesthetically pleasing diving experience while saturated in Hydrolab, I also had the most dangerous. It was the only time I ever lost track of the quantity of air in my scuba tanks. We were working on the sand bottom, making observations on garden eels one hundred feet below the surface and a couple of hundred yards from the habitat. I checked my gauge and was shocked to see I was almost out of air. Fortunately, I was working with Myra Shulman at the time, and being smaller than me, she consumed less air and had plenty left in her tanks. I drew a flattened hand across my throat, the universal sign that I needed air, and she handed me her octopus, as the secondary regulator is called. We returned to the habitat, swimming in tandem, attached by the air hose from her tank. Embarrassed to have made such a dangerous error, I didn't tell anyone about the episode, nor did Myra. For that, I was grateful to her at the time, but thinking back I see that my ego got in the way of good practice. To keep the managers fully informed to maximize safety, I really should have reported that my air supply had reached a dangerously low level.

In normal scuba diving, you immediately surface when something goes wrong, but this wasn't an option at Hydrolab. Our blood was saturated with nitrogen, and if we left the depths, symptoms of the bends could start within ten minutes of surfacing. Safety meant staying below until fully decompressing over twelve hours. An exception occurred when, on our first mission, Jim Tyler became ill. Normally gregarious and talkative, he lay quietly in his bunk, clearly in pain. The diving doctor on the support team made a house call to the habitat, diagnosed the problem as diverticulitis, and decided Jim needed to be evacuated. After careful planning, the support crew escorted him to the surface, speed-boated him to the shore base, and got him into a decompression chamber and safely repressurized to the equivalent of fifty feet—all in ten minutes. Following the long decompression period, a crew member took him to St. Croix General Hospital, but the staff decided he needed to be back in the US mainland to receive the best care. The hospital physicians would not release him for the flight unless a doctor accompanied him, so he traveled back with Dr. C. Lavett Smith. Smitty was not an M.D.—he had a Ph.D. in ichthyology—but nobody asked.

Normal decompression was a more methodical process. After a week below, we sealed the hatch, and over the course of twelve hours, we reduced the pressure in Hydrolab until it was equal to the surface air pressure. This was done manually. During their two-hour shift, each crew member sat in front of a foot-wide pressure gauge and a printed sheet of the pressures to be reached each quarter-hour. The on-watch aquanaut would slowly turn the large black knob controlling the outflow valve and similarly close the inflow valve to accomplish this goal. We had to ensure adequate airflow to avoid asphyxiation, and the critical nature of this task compelled us to check the pressure and flow gauges frequently, almost continuously.

Once the surface pressure was reached, we exited one at a time. When my turn came, I picked up a pony bottle (a small tank of compressed air) to breathe from on the journey to the surface, slipped into the trunk, and closed the upper hatch. I unlatched the lower hatch, but it remained sealed, held in place by the greater pressure on the outside. When ready, I gave the okay sign through a small port in the side of the trunk, and a support diver inside the habitat opened a valve that pressurized the inside of the trunk. I heard a great hissing sound, felt the pressure in my ears, and had to clear them several times. When the pressure matched that of the outside, the lower hatch fell open under its own weight. I slid through the hatch, and two divers escorted me to the surface.

Earlier, when fifty feet below, I had swum without scuba gear between the habitat and the tank rack, and I had a sense of complete freedom. During this ascent, I had zero freedom. I was instructed to remain completely passive, so I became a limp load while a diver on each side of me held an arm and slowly, very slowly, guided me to the surface at one-half foot per second. With complete trust in the support divers, I relaxed and watched the dark blue of the deep gradually become lighter and brighter as less and less water filtered the sunlight.

When my head broke through the surface, my ears were bombarded with splashing sounds, my eyes squinted in the unaccustomed brightness of the tropical sun, and I bounced up and down with the waves. I felt the pressure of the breeze and the warmth of the sun on my skin again. I was overwhelmed. I hadn't realized how sensorily deprived I had been. I imagine this experience of surfacing after a week below was very much like the one I encountered emerging from the womb, but I don't remember that.

Back at the shore base, we had chores to do and reports to write. The support crew was amused by the four of us because, apparently, we moved around as a tight little unit like a school of frightened fish. In the close confines of Hydrolab, we could not move unless someone else moved to make space, so our movements had become choreographed. At some level, we had become one.

One. Yes, we bonded and functioned as a well-oiled machine. While all our energy was focused on our research and daily chores, we did not give much thought to the surface crew. We were the Aquanauts, the stars, but also the amateurs to saturation diving. The support crew, the professionals, received less attention. Without their expertise, we amounted to nothing in this sphere. This disproportionate attention to the superficial players occurs in many fields, of course. Unsung support teams undergird politicians, doctors, architects, movie producers, city planners, and a host of other celebrated individuals across our culture. Without them, we are nonentities. I wish now I had thanked our support crew more fully than I did.

None of this was on my mind, however, as I glided over the white sand bottom in deep blue light fifty feet below the surface. And that, too, is what I recall now, beyond the scientific eurekas and frustrations, beyond the friendships and squabbles. That sense of floating alone in a different world . . . relaxed, safe, and remarkably unencumbered.

My Hippie Summer

In the summer of 1972, looking for adventure, I took a camping trip—my first break in five years of graduate school. Not a minor thing, I traveled with four companions for two months in a converted van through the American South, Mexico, and Belize. We chose to visit Belize after we had read an article in National Geographic[20] about the new country, formerly British Honduras. On returning from the trip in August, we sold the van, tallied up our expenses, and found that the expedition cost less than had we stayed put and paid our normal living expenses.

The group consisted of me, Floyd, Karen, her younger brother Glenn, and Janet, a graduate student. We purchased a used Dodge van powered by the indestructible Chrysler slant-six engine. It gave us no trouble. We built an elevated floor in the rear compartment to create storage space, purchased assorted camping equipment from army-navy surplus stores, stocked up on canned foods, and hit the road, leaving New Haven on a rainy day in June.

On our first night, we set up camp in the Blue Ridge Mountains in Virginia. We slept in jungle hammocks strung between trees. These were thick canvas hammocks with waterproof nylon roofs and mosquito netting sides. We had barely settled in when a strange sound pierced the night: "Whoa, whoo, whoo whoa, whoo, whoo, whoa." Glenn had spun out of control in his hammock, and we found him wrapped in the bottom, sides, and roof, suspended face down. He looked like a fly wrapped in spider silk, waiting for the inevitable. After a long laugh, we untangled him, at which point he moved his sleeping bag to the roof of the van, where he slept for the duration of the trip.

Floyd and I wore large beards, and Glenn had long, stringy hair—this was well into the hippie era. Our second stop was in an Alabama campground. It was a Saturday. In the evening, several wholesome-looking young girls walked among the tents, encouraging campers to attend church services the next morning. They were handing out pamphlets; one was entitled "Did Jesus Have Long Hair?" and made the argument that he didn't. Needless to say, we did not attend the services.

On through Mexico, where we could not believe the wonderful sweetness of field-ripened pineapples that we ate until we tired of them. We crossed many small bridges, each with toll collectors demanding "ocho pesos." I still don't know if they were legitimate officials, but the locals did not pay nearly as much, if anything at all. We were "poor" graduate students but recognized we were wealthy compared to them, so we did not resent the demands.

Being on the road in Mexico was quite an experience. At night, there were few cars, and we were told to be careful because sometimes black cattle would lie on the road. They'd be impossible to see in time, so we avoided nighttime driving. One evening in the Yucatan, we could not find a campground and were reduced to sleeping in a field beside the road populated by trucks transporting goods. These trucks had no mufflers and made an awful racket. We slept fitfully, worried about our safety and aroused by each passing truck.

There was little attention to road safety at the time when it came to signage, vehicle maintenance, and other measures. We came upon an auto accident in the rain on a twisty mountain road and stopped to help injured people until medical help arrived. To keep them warm, we used our sleeping bags, which ended up soaking wet and covered in blood. A few days later, on another narrow mountain road, we passed a truck on its side, its load of bananas strewn down the hill. Fortunately, we got through Mexico unscathed.

We eventually reached the Belize border. There were no paved roads in the country back then, and we proceeded down the Great Northern Highway, kicking up dust and rounding islands of trees in the middle of the road. Belize City was a bustling town full of friendly people at the time; it has since become a dangerous place filled with drug dealers and has the third-highest murder rate in the World.[21] The city is barely above sea level, and we were struck by the sewage system. Small water-filled ditches ran behind the houses, and each house had a rough wooden outhouse perched over the ditch. There was hardly any water flow, so each ditch was a stinking soup of human waste baking in the intense tropical sun. These ditches eventually drained into the Belize River. Robert Trench, a professor of mine at Yale, had told us about growing up in Belize City and swimming in the river. His semi-facetious comment was that his friends who did not get sick and die were immune to virtually everything the environment could throw at them.

The river was lined with fishing boats. These were twenty-five-foot wooden sailing vessels with open decks and little half cabins. We hired one to transport us out to Goff's Caye on the barrier reef. The fishermen were headed further out to Turneffe Atoll to find lobsters. We arranged for a second fishing boat to pick us up a week later and return us to Belize City. This may sound a bit reckless, but we registered our plans with the British military office, and they agreed to check on us if we did not report back in a week. The British were still there after independence to prevent intrusion by Guatemala, which claimed some of the western parts of Belize.

Unloading at Goff's Caye after sunset. Photo by author.

We loaded our camping supplies into the boat and left quayside in the midafternoon. The three-member crew immediately lit up their marijuana cigarettes, and we now knew how they passed the long, empty hours sailing back and forth—they had big smiles on their faces. We reached Goff's Caye as the sun was setting. It was a bare sand island less than half a football field in area with eight or ten short coconut palms and a ground cover of beach morning glory vines.

In unloading our gear, we had inadvertently left some of our water jugs in the dark hold of the boat and did not discover this until the boat had disappeared. We were stranded on a desert island with inadequate water for a week. Not to worry. We came up with two solutions to get us through. One was to use the large canvas tarpaulin, our shelter, to catch water from the frequent rain squalls. This method worked, but the water had an unpleasant taste from the anti-fungal chemicals embedded in the fabric. The second solution was to dig a hole in the sand in the middle of the island. I did that and found the water to be brackish but drinkable if it got to that point, and certainly fine for cooking. However, when I checked on the hole the next morning, it was empty, and then later in the afternoon, it was full again. That's when I realized that I had tapped into a freshwater lens in the sand. The freshwater layer floated on the denser seawater penetrating the island and it was rising and falling with the tides.

This was not the only lesson we learned on the island. I speared a large grouper, but it was too much for us to consume fully in one supper. We ate half of it and left the remainder out, thinking it would still be fresh enough for breakfast. However, when the sun went down and the stars blazed above, there was another source of light on the island—the grouper was glowing. That turned us off the fish, and we threw it back into the ocean for the crabs, shrimp, worms, and other scavengers. I have since learned that the light emanated from luminescent bacteria that require oxygen, so as long as the fish glowed, it was fresh enough to eat.

Above. Camp at Goff's Caye, Belize. Photo by author.

Left. The luminescent grouper with red handle of knife and green spearhead. Goffs Caye, Belize 1972. Photo by Floyd Connor.

Partway through our stay, a motorboat pulled up onto the island and out jumped two men, one stout and heavily bearded. He was David Stoddart, an eminent physical geographer from the University of Cambridge who was mapping the vegetation on the islands of Belize, documenting changes after Hurricane Hattie. We spoke for a few minutes before they walked the island, taking notes on their clipboards, then were gone, off to the next island. Here I was, in Belize, before I started my research career on coral reef fishes, and by serendipity, I met one of the future giants in coral reef research. I didn't know who he was at the time. He went on to become co-founder and first president of the International Society for Reef Studies. I ran into him again thirteen years later at the Fifth International Coral Reef Congress in Tahiti but failed to mention our meeting on Goff's Caye.

Sarah Lawrence College

My education was very focused—science, science, and more science. I took as much as I was allowed as an undergraduate, and of course, graduate school was all science all the time. I applied for a professorship to several universities in the US and Canada and made the shortlist at four. The shortlist comprises three or four candidates invited to visit the campus and deliver a seminar. Making the shortlist out of a couple of hundred applications is an accomplishment but meaningless if you're not The One. Instead of a traditional research/teaching university position, I made the final cut in 1972 at a small liberal arts college with an emphasis on the humanities and the arts, Sarah Lawrence (SLC). I had to make a transition, and that's quite an understatement. Instead of being surrounded by like-minded colleagues who thought the way I did and understood why I studied the fishes I did, I was part of a community of scholars who approached their academic interests in an entirely different manner. I had nobody to run my ideas by. I sat at lunch, listening to conversations about things I didn't know. But my colleagues were mostly kind and supportive. Around 1990, the Internet emerged. Websites and email became widely used, and I was no longer isolated. I could easily communicate with old colleagues and make contact with new ones.

But teaching—I had not done that before. That scared boy who had put his arm in the Bunsen burner flame in front of the class back in high school was still there. I froze giving a seminar in graduate school, and the professor kindly suggested I present another day. I managed the second time. Teaching a college class was a major stress for me. I prepared very detailed notes and presented them in a stilted manner. Once, I actually stood in front of a class, with fifteen faces looking expectantly at me, and thought: "*Oh shit, here I am standing in front of a class.*" It took a few years before I became relaxed and a better presenter, and I think I became quite an effective teacher. Still, if someone asked me if I preferred to go out in front of a class or to work at my desk in the office, I would choose the latter.

General Biology, Ecology, Marine Biology, Evolution, Human Evolution, and Environmental Studies were the regular courses I cycled through. I also taught a variety of more specialized courses from time to time, including Coral Reefs and Oceans in Peril. The system at Sarah Lawrence requires that, in each of their courses, the students meet individually biweekly with their faculty to carry out large projects culminating in significant papers. Therefore, most classes are limited to fifteen students, and faculty teach two courses at a time. The range of project topics in a given course could be very broad. Most of the students I taught were not science majors but came to me with a wide range of interests. I always made an effort to find something each student could relate to and get excited about.

In an Evolution course, for example, as I was thrashing out project ideas with a theater student, we settled on a comparison of ape behavior with human behavior. Her ingenious approach was to take the published account of the dynamics of a power struggle and takeover of dominant male status in a chimpanzee troop at the Arnhem Zoo (*Chimpanzee Politics*[22] by Franz de Waal) and compare it with the plot line of Shakespeare's *Richard the Third*.[23] In each, she found an alliance to overcome the current leader and a betrayal after success. In fact, she was able to show that the roles of specific characters in Shakespeare's play mirrored the roles of specific members of the chimp troop—or was it the other way around? Based on the fact that we share 98.6% of our DNA, Jared Diamond, an ecologist at UCLA, called us the "third chimpanzee" (along with the chimp and bonobo).[24] The theater student illustrated that we share much social behavior with them as well. For example, they will not cooperate with cheaters. This basic continuity is central to evolutionary theory. We are basically animals with a cultural veneer, but essentially animals nevertheless.

Student projects could also provide valuable contributions to faculty research. Later on in this book, I describe how a student doing a project for me provided an insight that informed all my subsequent work. On another occasion, I benefitted from a student's project for a colleague. To catch the fish I studied, I used quinaldine. It is a nasty reddish-brown, smelly, oily liquid derived from coal tar that happens to be a quick-acting fish anesthetic. It's not very soluble in water, so when I first started using it, I made a solution in ethyl alcohol. However, quinaldine is noxious, can cause skin irritation, and slowly degrades unless stored in an oxygen-free atmosphere.

In 1977, I read a paper[25] describing a method of converting this nasty liquid into a stable, water-soluble white crystal that looks a little like sugar. The method required reacting the quinaldine with concentrated sulfuric acid and going through a complex purifying process. So, I turned for help to Dave Brewer, my brilliant friend and colleague in the Chemistry Department. He said this synthesis would be an ideal student project. He enlisted a student in his Organic Chemistry class, and she worked hard in the lab, gifting me at the end of the semester with a few ounces of quinaldine sulfate, a lifetime supply. Being a pure, dry crystal, it has a very long shelf life.

During graduation week, it was traditional for the faculty to put on a show for the seniors, the most receptive audience imaginable. Some sang, some played instruments, and some acted in skits. Not being talented in that way, I created a different performance. I began the project by going to the library and borrowing *The Social Behavior of Animals*[26] by Nikolaas Tinbergen. In it, he describes the mating behavior of the three-spined stickleback, a little fish widely distributed in ponds and seashores both in Europe and North America. The male builds a nest out of algal strands, and as groups of drab females swim by, he tries to attract one to his nest by displaying his bright red belly in a zigzag dance that ends with him poking his snout into the entrance of the nest.

Working with several other faculty members, one a choreographer from the dance department, I produced a dance displaying this behavior. The performance started with me alone at the front of the stage in full scuba gear. I removed my mask with a flourish and began reading from Tinbergen's dry technical account. One by one, faculty appeared behind me, ostensibly without my knowledge, dancing out the piece I was obliviously reading. The male strutted around while a bevy of females skittered about until two clasped hands above their heads creating an arch, the nest. The climax came when Bill Park, a senior member of the literature faculty, tore open his shirt to expose the red tee shirt beneath it, just as I read the part about the male displaying his red belly. Not very subtle, but the crowd went wild.

Over the years, I learned a great deal and actually got my informal liberal arts education while I was a faculty member. Importantly, I learned not only from my faculty colleagues but from my student colleagues as well. Our conversations about their other courses and major interests opened whole new worlds to me. I may not be an art aficionado or a connoisseur of the great novels or a font of philosophical theories, but I at least have some familiarity with them, a familiarity that came neither from my home life growing up nor my formal education. I am now a strong advocate of the liberal arts and believe they are the best preparation for responsible citizenship. I also believe that science courses are an essential component of a fully balanced liberal arts education.

The liberal arts, and I include pure science here, have frequently been accused of having no practical value. I've always resented hearing people contrast academia with the "Real World." I would hear things like: "Maybe it's like that in college, but in the Real World..." Some of my colleagues would also speak this way, but the academic world is every bit as real as any other world. In fact, I could make the case that the commercial/financial world is not any more real than the academic world. Money, especially on the Internet, is not a tangible thing, and its value is arbitrary and changes over time. From that perspective, it's farmers who live in the Real World, producing something that people need no matter what economic system they live in.

My Fascination

St. Croix, 1979. That tiny head was still there, just like yesterday and the day before. It was peering out of a hole in the coral, its bright yellow googly eyes moving independently as they scanned the water. It was a fish, a fish like no other I knew, in a study plot I was monitoring to determine the diversity of fishes that pass through a square yard of reef. I had to name it to add to my growing species list, but I needed to have it in hand to make the identification. Not wanting to interfere with my study plot, one morning I removed a similar fish from another location to take back to the lab for identification. It was a spinyhead blenny. As I passed by the emptied hole on my return to the study site in the afternoon, I checked it and saw that another individual had taken up residence in it. I removed that one, too, and the next day another fish was there, and again the day after.

A spinyhead blenny peeking out of its home in an abandoned worm tube. Photo from https://www.costambarbeach.com/blenny-fish/

I began to wonder if there was a floating population of blennies without holes ready to pounce when an opening occurred. This idea came from my previous reading as a student. In a classic ecological study in the 1950s,[27] Robert MacArthur removed the territorial male warblers in a spruce forest in Maine and found that they were quickly replaced. He concluded that there was not enough space for all males to hold a territory and that a floating population of non-territorial males was constantly watching for open territories to occupy. Over time, I learned that all blennies do have holes to live in, but they are always looking for better shelters, very much like hermit crabs that are always looking for better shells to live in.

Unlike most fish that maintain neutral buoyancy by having gas-filled swim bladders—enabling them to stop swimming and remain motionless, neither sinking nor floating—blennies are denser than water. When not swimming, they sink. Spinyheads belong to the tube blenny family along with the roughhead blenny. These are little fish about an inch in length, long and narrow but shorter than their genus name, *Acanthemblemaria*. I know these species intimately.

By turning my gaze to a close-up inspection of the reef surface, I saw that the blennies I had previously overlooked were quite abundant. I was hooked, and off I went into thirty years of blenny study, although I did not know it at the time. This proved to be a fortuitous choice for another reason. Few investigators were focusing on the ecology of these blennies, so I had the field mostly to myself. I was at Sarah Lawrence with a very demanding teaching load, so I could not engage in research with the same intensity as biologists at research universities. I could work at a manageable pace and without being "scooped." But why was this group of fishes receiving so little attention, and more importantly, was it worth my time investigating them?

Ideal Model Organisms

Have you wondered why geneticists study fruit flies? They are not intrinsically more interesting than other animals. Newly laid fruit fly eggs will become adult egg-laying fruit flies in three weeks. Because large numbers can be maintained in little glass bottles, many colonies can be kept in a small space, sustained by nothing more than bananas and yeast. They have a great number of genetic variants that are easily recognized. Consequently, pure strains of fruit flies can be quickly bred and crossed with other strains to produce numerous hybrid offspring in a short amount of time. And, since the basic principles of genetics are common to all animals, fruit flies can be used to efficiently elucidate the general patterns of inheritance. They are ideal model organisms.

The patterns of ecology are much more variable than those of genetics, so no animal can be as universal a model as the fruit fly, but some regularities in biological communities apply quite widely. So, even though they have no food or other commercial value, I found blennies to be very suitable fish to use as models in understanding the precise ways in which reef fishes partition resources and thus avoid competition. Blennies are very poor swimmers intimately tied to their shelter cavities. Most fish are mobile and will not remain in locations to which they are transplanted, so certain experiments on fine-scale use of space are usually impossible to perform. Blennies, however, can be moved to new locations and will stay there. That makes them excellent subjects for field experiments.

Anesthetized blenny drifting into a test tube. Photo by Chris Finelli

Furthermore, whereas most fishes are difficult to capture as they quickly slip among the branches of corals, blennies withdraw into their holes when threatened. All I need to do is use a syringe to inject a little quick-acting anesthetic into a hole, place a test tube over the entrance, and in less than ten seconds, the blenny slips into the test tube, out cold. Virtually every fish I target is captured, an unheard-of success rate among reef fishes, and they quickly recover from the anesthetic without any after-effects. So, like fruit flies, blennies are well–suited as model organisms.

Later on in my career, I needed to mark individual blennies so that I could investigate their homing ability. At the time, a new technique for marking small fish was to inject acrylic paint under their skin with a hypodermic needle. With four anatomical locations and a wide range of colors, many individuals could be uniquely marked. I prided myself on the skill I developed. Thirty feet down, in full scuba gear, while being pushed back and forth by the ocean swells, I could inject a streak of color under the skin of an anesthetized inch-long fish. Before any of that could happen, though, I had to acquire materials. Syringes and needles came from the stock supply in the lab at Sarah Lawrence, no questions asked. The paint? Well, that led to an awkward encounter with the clerk of an art supply store. How do you get a non-biologist to understand why you want to inject paint into live fish?

Blennies are also easy to maintain in aquaria and can be used in laboratory experiments as well. Back in the day, I kept a dozen in a small tank on my kitchen counter for four years. On many flights, I successfully transported sixty blennies in plastic bags in an insulated box that was my carry-on luggage. This became a little problematic in 2007 when the Transport Security Administration's new rules limited carry-on liquids to 3 ounces. I once went through the inspection station at the Houston airport, hoping the supervisor would be reasonable and understand that fish could not live in bomb-making chemicals. When he looked at the plastic bags, he said they were all dead. I had to explain that these fish normally sit on the bottom rather than swim in the water. Reason prevailed, and I was home free. On the other hand, the yard-long pieces of steel rebar that I used to delineate quadrats for sampling were confiscated—potential weapons, you know.

Habitat and Behaviors

Spinyheads and roughheads reside in cavities in living and dead coral. They choose tubes that closely match their size and spend most of their time with their heads poking out as they scan the surroundings for food. They eat a wide variety of tiny animals, but the bulk consists of crustaceans called copepods, each smaller than a grain of rice. Rarely leaving their cavities, blennies are virtually fixed in place. Much like the attached barnacles, they depend on water movement to deliver their food to them. Whereas barnacles sweep the water with their net-like appendages, blennies select individual items. Their large eyes bug out, each moving independently as they follow passing organisms. When they see a tasty object, they shoot out of their holes, grab it, and zip back in a flash. They spend most of their day watching and eating—not how I would choose to spend my time.

Clearly, this cannot be all they do, I thought. From an evolutionary perspective, their lives are meaningless unless they reproduce, which I discovered happens around sunrise. Ripe females leave their cavities and scoot along the coral surfaces, attracting the attention of males who begin courting. Courtship involves the males rapidly thrusting their whole bodies in and out of their holes in a staccato rhythm. At times, five or six males may be directing their attention to the same female. After some consideration, she chooses a male, enters his cavity, lays her eggs, and leaves. He fertilizes the eggs and spends three weeks fanning them as they develop. Fish dads, in general, are superior caregivers than we mammals. In fish, it's usually the male that nurtures the eggs, whereas mammalian males seldom contribute more than their sperm. To my surprise, some male fish actually match female mammals in embryo care. Camilla Whittington of the University of Sydney has recently shown that male seahorses actually build a placental connection with the embryos that are developing in their pouches.[28]

I don't know how the female blenny makes her choice. She may do more than simply evaluate the hole's location and the quality of the male's "dance." An interesting study[29] by Sarah Kraak, then at the University of Leicester, showed that females of the sphinx blenny in the

Mediterranean Sea lay a few test eggs in the holes of several males and come back the next day to determine the survival rate of those eggs. They choose the most successful male and lay the bulk of their eggs in his cavity. The more carefully we study these fish, the more sophisticated they prove to be in all the choices they make. I cannot exclude similar sophistication from spinyheads and roughheads. Nobody has looked at that.

Although they generally remain fixed in their tubes, blennies are constantly watching each other and are prepared to switch if a fish in a superior hole vacates it. The immediacy of this switching was impressed upon me when I was catching blennies for an experiment. I squirted some anesthetic into a hole and placed a test tube over the opening. As soon as the resident came drifting out, a neighbor zipped over and banged against the side of the test tube as it tried to occupy the hole.

Sometimes, they don't wait for a resident to leave. Several times, I have seen a blenny approach another in its hole. When the resident withdrew, the aggressor stuck its head into the hole and began thrashing its tail and twisting like a corkscrew, evidently having a bite on the resident. It then withdrew and waited by the opening. The resident quickly swam away, and the aggressor occupied the hole. This violent behavior was even more obvious when I was placing blennies on an artificial habitat for an experiment. I observed a blenny in an occupied hole defend it against another by grabbing the latter in its jaws and holding it up off the surface for a few seconds before releasing it. These guys may be small, but they can be major bullies in their world.

Where to Begin with Blennies?

Field research involves close observation of behavioral and ecological phenomena. In its modern form it is fancied up natural history, the detailed description of the living conditions of plants and animals in their natural environments. We look for patterns and similarities between different groups of organisms with similar characteristics to find generalities, or theories, about causal relationships. We first express these as tentative statements—hypotheses—to be tested.

So, most modern field biology involves hypothesis testing. Some testing is done through systematic observations of unaltered nature, some through manipulative experiments. Most people think of experiments as complicated procedures that occur in the laboratory where all variables are controlled, but experiments can also be performed in the field, and they can be quite simple. By systematically controlling only one variable while the rest fluctuate in their natural ways, we can separate the controlled variable from all the others and attribute the organisms' response to that one cause. Furthermore, because we are performing this exercise in the normal habitat, the results are more likely to be meaningful to the biology of the organism in the natural context.

I am constantly amazed at the ingenuity that many investigators show in the design of their field experiments. For example, Malte Andersson demonstrated that the long tails of male African widowbirds are the result of female preference.[30] He did this by cutting the tips off the tails of some, using cyanoacrylate glue to attach the removed lengths of feathers to the tails of others, and using unaltered males as a control. He found that males with the shortest tails had the lowest mating success, males with the longest tails had the highest success, and males with intermediate length tails had intermediate success—elegant experiment, clean results.

As I mentioned, roughheads and spinyheads make excellent material for studying behavioral aspects of reproduction and competition. Initially, though, I did not do those kinds of studies. My interest was at the ecological level. I wanted to explore the mechanisms behind the incredible species diversity of coral reef fishes. Among the various hypotheses is the idea that fish species specialize on different resources and, by so doing, avoid competition. Why was that my specific interest? It may have been the influence of my teacher. I was introduced to the concept in 1965 when I took my first ecology course in my junior year at McGill University. My professor, Peter Grant, was specifically interested in this topic, and I also found it fascinating. I think it's the potentially clean explanation of a complex phenomenon that attracted me. It turned out that the real world is not so simple.

Resource Partitioning

Part of the beauty of the natural world lies in its diversity. There are so many species, and you are often confronted with the unexpected in many incredible ways. This diversity also applies to the chemical and physical aspects of the environment in which plants and animals are embedded. But complexity, for all its attractiveness, presents a problem to the scientist. While we can describe patterns of distribution and behavior, we run into difficulty in attributing causes because, as you move from place to place, many variables change simultaneously. For example, you may encounter different current speeds, different light intensities, different corals, different algae, different snails, and an endless number of other differences. If spinyheads and roughheads are found in different locations, how do you know which of so many varying features is the cause of this separation? Perform a field experiment.

In the early 1980s, I noticed that spinyheads were situated high up in the corals, while roughheads were found at the bases of corals and in the carbonate pavements to which the corals are attached. In my surveys of blenny density, I measured the height above the pavement of each individual with a yardstick. The pattern proved to be very clear and persisted year after year. What factors caused this distribution? Did the tubes the blennies live in differ in low and high locations? The cavities can be very complex and difficult to characterize. Did spinyheads simply prefer being high, and roughheads simply prefer being low? Or did they both prefer the same locations, but one displaced the other from that location? Attempting to find answers, I designed an artificial habitat that provided a uniform surface with identical cavities at constant densities at different heights above the reef pavement.

The habitat was a concrete slab a little over a yard square and about two inches thick. One surface had a forty-by-forty matrix of cavities (1,600 total), each three-sixteenths of an inch in diameter and one inch deep. I made it by building a wooden form. The mold for the cavities consisted of a piece of pegboard with nails inserted into all the holes. A piece of plywood covered the side with the nail heads, holding them in place. It was then that I realized I had constructed the proverbial bed of nails. I made the concrete, poured it into the form, and placed the mold with the nails resting on the concrete surface. But when I pushed the mold down, it wouldn't go. I was stranded with a wet concrete-filled form and 1,600 nails that were not slipping into the concrete as I had expected. What to do? I went around the labs and offices and enlisted three people to help me. So, four Ph.D.s did a little dance, holding on to each other on a yard-square dance floor until our combined weight forced the nails into the concrete. After the concrete cured, I removed the mold and was greeted by a beautiful smooth surface with 1,600 perfectly formed, equally spaced cavities.

The concrete habitat, thirty foot depth at base of Teague Bay Reef, St. Croix. Photo by author.

The completed habitat weighed three hundred pounds. Using a boat with a heavy-duty swinging arm and winch, two members of the lab staff slowly lowered the habitat to a sandy bottom at the base of the reef thirty feet below. I was waiting down there, watching this behemoth descend. The boat was rising and falling with the waves, and so was the concrete slab at the end of the rope. I had to be very careful to avoid being hit in its herky-jerky approach. Wrestling a bouncing three hundred pounds underwater took a lot of strength, and I started breathing heavily. Once I released the slab from the rope and had it resting on the bottom, I checked my air gauge—I was almost out of air! Until then, I had been hardly moving while counting blennies and was used to my air supply lasting up to an hour and a half. With the heavy physical work, I blew through my air in about twenty minutes. I immediately surfaced at the prescribed safe rate of one foot per second and felt lucky to avoid a more dangerous situation.

When finally installed, the habitat stood vertically on a sandy floor. I had recreated the monolith in Stanley Kubrick's movie *2001: A Space Odyssey*. Over the next several weeks, I captured blennies on the reef and transplanted them to the concrete habitat. When I placed spinyheads alone on the habitat, they assorted themselves with the greatest density on the highest part, replicating their distribution on the reef. When I placed roughheads alone on the habitat, they, too, assorted themselves with the greatest density on the highest part, contrary to their natural occurrence in the lowest locations on the reef. So, both prefer high locations.

In the natural habitat, did spinyheads displace roughheads from their preferred positions? I answered this question by placing spinyheads and roughheads together on the artificial habitat and found that the pattern on the reef was replicated: spinyheads were in the high locations, and roughheads were in the low locations. By creating this habitat in which all cavities were identical, as was the nature of the surface, the only variable in play was height above the substrate. I concluded that spinyheads competitively excluded roughheads from the microhabitat that both species preferred.[31] In science, it's become cliché to say that every question answered raises more questions, but it's the truth. So, I now wanted to know: what about height was so attractive?

Top: Plankton net being hand-towed.
Bottom: Sample jars from low and high tows (see text).
Photos by Chip Clark (top) and author (bottom).

Many of the copepods blennies eat live in the water above the bottom. I knew that other organisms—fish, corals, shrimps, and others—also feed on such copepods. This predation would result in a reduced density of copepods close to the reef surface, so being higher up in the corals should place the blennies in a location with higher food density. But was this really the case?

I answered this question by using a plankton net, which is a long, conical, fine-meshed net open at the wide end and with a cup-like container at the narrow end. These nets are usually towed behind boats, but I was interested in fine-level distribution, so I swam the net over the reef at one of two levels—just at the reef surface and a yard higher. I found the pattern I had predicted: a group of planktonic copepods (calanoids) were always more abundant in the high tows than the low tows. In fact, during one period when there was an unusually high density of calanoids, their density in the high tow was one hundred times greater than in the low tow only a yard away. I was shocked that such an extreme difference could occur over such a short distance. This spectacular layering could not be caused simply by different predation rates. Clearly, the copepods were avoiding the bottom; they were not the passive drifting plankton they were made out to be. I could have started down that road, working to explain this pattern, but I needed to stay focused. The important point is that there was a difference in density and that the blennies were responding to it.

I was not satisfied by simply showing that the preferred position was the one with more food. Were blennies more successful there? I needed to do another experiment to measure the success of spinyheads and roughheads living at different heights. In 1988, I created six new habitats. Each consisted of two opaque brown acrylic plastic panels a quarter-inch thick and twelve inches high by twenty-eight inches wide. I drilled eighteen holes in each. Into these holes, I glued plastic screw caps—each cap had a three-sixteenths inch hole in it. I screwed small glass vials into the caps so that they projected from one side of the acrylic panel. I covered them with black rubber tubing so that they would be dark inside. The other side of the panel looked like a smooth surface with eighteen perfect blenny shelters in it. I attached pairs of these panels, oriented vertically, onto two steel rods, one panel on the reef surface and the other a yard above.

Experimental habitat used to determine blenny fitness in highand low locations. Photo by author.

I captured blennies on the reef and placed nine on each panel, placing the same species on each high-low pair so that I could compare how well the blennies living high and low in the same location would do. To measure the feeding rate, Karen, who generously helped me with my fieldwork, counted the number of attacks on prey occurring on each panel over a ten-minute period. At the same time, I removed the black rubber sleeves covering the vials on another panel and counted the eggs in each. I also measured the length of each fish at the start and end of the twenty-four-day trial.

The results were clear: both species attacked prey more frequently, grew faster, and laid more eggs on the higher panels.[32] Wow! Getting such clean results in what seemed like a long-shot experiment was so gratifying. I now knew the advantage of being higher up. Essentially, higher blennies were better nourished and thus left more offspring. In Darwinian terms, they had a higher fitness. It was inevitable that the blennies would evolve a preference for high locations. But are they eating the same things as they occupy their characteristic locations? Are roughheads simply getting less of the same calanoids that spinyheads seem to be eating? I could not fully understand their relationship until I knew what each species ate.

The only way to get at the question of diet was to examine the stomach contents, and that meant killing the fish. I didn't like doing that, but there was no alternative. Because I needed the stomach contents to be as fresh as possible, I preserved the fish in formaldehyde as soon as I got into the boat. That wasn't always as easy as it may sound. I would be on my knees on the bottom of the boat, still in my wetsuit, trying to transfer fish and fluids while everything was rocking in the waves.

To get a handle on the degree of selectivity in diet, I needed to know the stomach contents in relation to food availability in the environment. To this end, I used the precisely guided plankton tows described earlier. I also collected pieces of dead coral and coral rubble. These pieces were transported in individual plastic bags and preserved in formaldehyde. By vigorously shaking the bags and removing the coral pieces, I was able to obtain samples of the tiny animals living among the fine algal strands on the reef surfaces.

I now had my samples of potential prey living in the open water and on the surfaces. That was the easy part. What remained was the tedious chore of identifying all the animals in thirty surface samples, thirty plankton samples, and one hundred and fifty fish stomachs. As I said, teaching at Sarah Lawrence was a very demanding, time-consuming full-time job. For about a year, my routine was to end my week with Friday evening in the lab. This was not as onerous as one might expect; I looked forward to the discoveries I would make in each sample. I would be able to process two or three samples during each three-to-four-hour session. The plankton and surface samples were relatively easy; I would place them under a microscope, identify, and record each animal encountered.

The stomach samples were another matter. I had to dissect the stomach out of each inch-long fish and rinse the contents into a tiny dish. This is very precise work involving fine forceps and micro-scissors. In performing the task, I was struck by the remarkable adaptability of my nervous system. In executing any task involving our hands, our eyes give us feedback to coordinate our fingers. When I looked through the microscope, the movements of my hands were magnified thirty to fifty times. In other words, if I moved a needle one-sixteenth of an inch, my eyes would see it move apparently two-and-a-half inches in the microscope field of view. Somehow, my brain was able to adjust instantly and make the necessary tiny movements even though my eyes were seeing a view that would normally require much larger movements. How do we do that?

The other challenge of the stomach samples was that some of the organisms were partially digested, which made identification difficult, but after some time I had enough experience to identify animals from their disarticulated parts. This task was aided by a large collection of photographs of intact animals that I had taken through the microscope.

Once I had counted how many of each of the seventy-eight taxonomic groups occurred in each of the samples, I could compare what blennies were eating in high and low locations in relation to the prey availability in the water and on the surfaces in those locations. The comparison involved considerable analysis on the computer, including the application of mathematical indices of selectivity. Such indices provide numbers indicating whether and to what degree fish were eating more or fewer prey in relation to their abundance in the environment. I also enjoyed this work—it was a lot like puzzle-solving.

Sometimes the results slowly emerged, but sometimes there were "aha" moments when, after methodically setting up an analysis, I pressed the "return" button, and results instantly appeared on the screen. It turned out that spinyheads were eating primarily copepods that live in the water column (calanoids) and roughheads fed mostly on different copepods that live on surfaces (harpacticoids). Furthermore, this distinction did not simply reflect variation in the abundance of these copepod types in their locations. Spinyheads ate more calanoids and fewer harpacticoids than in their environment, and roughheads ate fewer calanoids and more harpacticoids than in their environment. In other words, they were specializing on the most abundant prey in their respective locations.[33]

Collaboration is Key

I still didn't know the mechanisms underlying the food specialization and, for that matter, what traits allowed spinyheads to displace roughheads from the preferred high locations. In an effort to solve the latter question, I brought live fish back from my field site in the Virgin Islands to my lab in Bronxville, New York. I built a tall aquarium with blenny tubes I could monitor and was ready to start the experiment. Travis Gering, my stellar undergraduate student, was set to do this project. Unfortunately, that did not work out, but he did make a key observation. One day, while we were discussing the project, he said: "You know, the gill flaps of spinyheads beat faster than those of roughheads."

That statement set off a chain reaction of hypotheses and field and lab studies lasting 18 more years. I thought: *Gills move faster. That means they're pumping water faster and getting more oxygen. They must have a higher metabolic rate.* Acting on that thought, I measured their metabolic rates,[34] which are indeed higher in spinyheads. From there, I inferred that spinyheads need more food. The water also moves faster higher up, delivering more food but requiring stronger swimming abilities. I also hypothesized that spinyheads, with their higher metabolic rates, would win fights and, therefore, the desirable higher sites in the coral. I arrived at this idea after reading a paper by Bernd Heinrich and George Bartholomew on African dung beetles.[35] It turns out that dung beetles fight over balls of elephant dung, and the beetles with the warmest body temperatures, and thus the highest metabolic rates, are most likely to win fights. A good naturalist does not wear blinders restricting himself to one group of organisms.

It's hard to imagine where my studies of blenny resource partitioning would have gone without the central determinant of metabolic difference. Undoubtedly, I would have uncovered it because it was there and had so many consequences. But it probably would have emerged slowly. I did not have the time to watch individual blennies up close as my student did, but maybe he was simply more perceptive than I. In any case, I can point to that moment as a transformation in my understanding of the relationship between these species.

Did the different metabolic rates of spinyheads and roughheads really influence their microhabitat distribution? The copepods that spinyheads consume live in the water column, and the copepods that roughheads eat live on surfaces. I wondered if faster spinyheads had an advantage over roughheads in capturing prey with escape mechanisms as they zipped by in the oscillating currents generated by waves.

I had for many years wanted to take close-up video of blennies catching their prey but did not have the technical expertise. One day in 1997, I read a news piece in *Science* magazine about Petra Lenz at the University of Hawaii, who took remarkable electron microscope images of copepods. I looked up her web page, and in it, she mentioned a coworker in Texas who was taking close-up video of copepod escape behavior. I Googled the coworker, Ed Buskey, at UT Austin and sent an email inquiry: would he be interested in making video shots of blennies eating copepods? He was indeed interested because he had been using a tiny vibrating sphere to stimulate escape behavior and would really like to use a real predator. The fact that blennies live in tubes makes them especially suitable for close-up imaging because the feeding events occur in predictable places. To our great fortune, we both had received grants for research trips to the Smithsonian's Carrie Bow Caye field station in Belize, so we arranged to be there at the same time to meet and evaluate the potential. Ed was really impressed with the ability of blennies to elicit copepod escapes in a predictable spot on which a close-up lens could be focused. We committed to working together, and eventually we published three papers. Thank goodness for the Internet. I would never have connected with him at a conference; we moved in different circles and attended different meetings.

I received a small National Science Foundation grant, which piggybacked on Ed's large grant, and we were off. I made a collecting trip to St. Croix and flew with a box of blennies to his lab at Port Aransas on the Texas coast, arriving after eleven at night. Ed had prepared sixteen small glass tanks with flowing seawater. Each tank had four small pieces of Sculpy modeling clay into which I had drilled a hole for a blenny to occupy. The tanks sat on a water table. Via a maze of clear plastic tubing, seawater was flowing into each tank and overflowing onto the table and out the drain hole. We placed four blennies in each tank and left the lab to get some sleep. I received a shock when I returned in the morning. Because blennies are bottom-oriented and we provided shelter holes, I thought they would not swim up and over the edge of the overflowing tanks—wrong! Some blennies were in the shelters, but many were out of the tanks and on the water table, barely covered by the quarter inch of flowing water. We captured them and returned them to the tanks with the water flow turned off. I cut pieces of fiberglass window screening and glued them to the top edge of each tank to solve the problem. I turned on the water, and disaster was averted.

Ed's lab was a combination of museum and parts store. The shelves were filled with a diverse array of glass tanks, rubber tubing, water pumps, and mysterious acrylic sculptures that had been parts of past experiments. The drawers were filled with wire, switches, LED lights, and a variety of electronic components. Surrounded by these riches, I set to work designing an apparatus in which to study blenny feeding under different conditions of water turbulence.

I constructed a little two-and-a-half-inch by five-and-a-half-inch tank with a built-in water pump. I could transfer a blenny in its shelter from a holding tank into this "turbulence tank" and could watch a blenny catching copepods under turbulent and still conditions (pump on, pump off). Ed set up cameras to capture the interactions. To get a good depth of field with close-up lenses, he needed very bright light, so Ed used infrared LEDs. The beauty of these is that neither fish nor copepods could see infrared wavelengths, but the camera could.

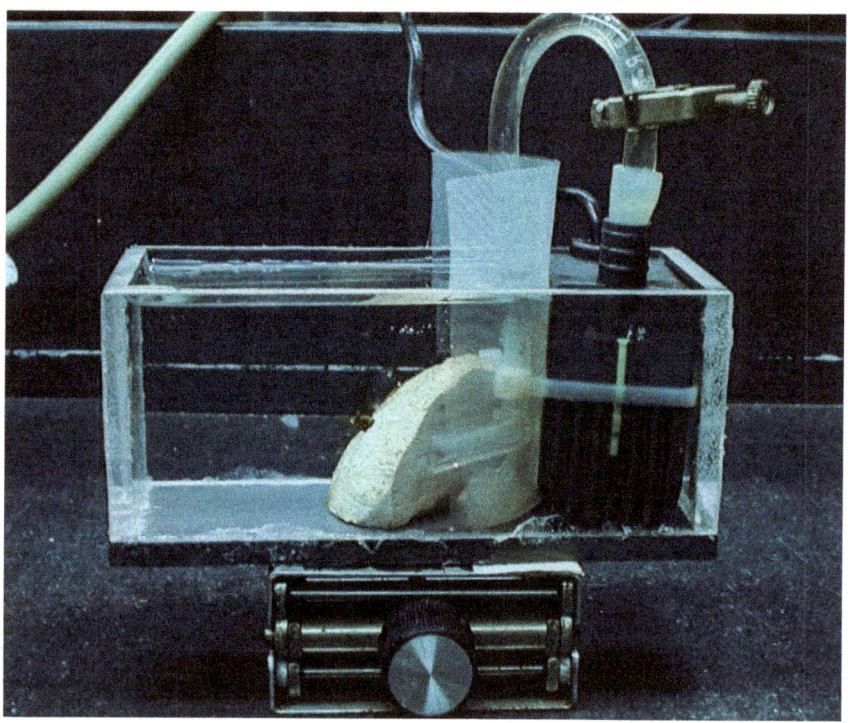

The first simple "turbulence tank." Photo by author

I should be clear here—I was studying blenny feeding, but Ed was studying copepod escape mechanisms. What I considered a successful capture, he considered a failed escape, and vice versa. It's not that we rooted for different outcomes like spectators at a sporting event. However, we were trying to answer different questions, and we had the great fortune to bring our various skills together to address our individual interests using the same experimental system. Such collaboration with different objectives is very rare.

Our results showed that spinyheads were more adept than roughheads at catching copepods in turbulent water but that roughheads, using a stealthier approach, were more successful under still conditions.[36] These results were consistent with our measurements showing that water flow was greater in the higher locations occupied by spinyheads, as well as with my demonstration of a higher metabolic rate in spinyheads.

Flow on the reef is caused by waves, and as a result, it oscillates with rapid changes in speed and direction. Aside from speed, the level of turbulence also affects the capture success of blennies. I felt we needed to create a more natural flow pattern for the lab study, and we also needed to know what that flow pattern was. Ed had met a young oceanographer at a conference and suggested we contact him to see if he was interested in collaborating with us. Chris Finelli joined our team and added a tremendous amount of technical capacity to our work. Chris's special expertise was in measuring water flow on a tiny scale in the natural environment. The three of us wrote a research proposal to the National Science Foundation, and we were fortunate enough to have it accepted, providing us with a budget to do the work.

Chris used a variety of instruments that, to me, were magical in their function. For instance, the "acoustic Doppler velocimeter" produced up to 16 million sound pulses per second and recorded the echoes reflected off the tiny particles naturally occurring in the water. From the change in these echoes, the speed and direction of the water were calculated in a volume much less than a teaspoon. As a result, we were able to measure water movement at blenny shelters to document the precise conditions in which the fish were feeding.

But how to recreate in the lab the speed, oscillation, and turbulence we measured on the reef? Here, too, Chris had the answer. He designed two flumes—enclosed tanks with controlled water flow. One flume produced one-directional water flow at varying speeds with the capacity to vary turbulence level. The second generated an oscillating water current at varying speeds, so the fish were exposed to water moving in one direction, slowing down and stopping, then increasing again to high flow in the other direction. He designed the flumes to recreate the flow speeds, turbulence levels, and "wave" periods to replicate the conditions we measured on the reef. Chris not

Wave-simulating flume. Blenny is in small white square at center. Photo by author.

only had the knowledge and creativity to design the flumes, he was also on the faculty of the Louisiana Universities Marine Consortium, which had the shop and skilled technicians to turn his drawings into functional flumes.

We spent two weeks during the summers from 2004 to 2006 conducting studies at the Middle Cay field station on Glover's Reef in Belize. We ended each expedition by capturing blennies and transporting them to Ed Buskey's lab in Port Aransas, Texas, where we conducted the flume studies. Each year, I chose a Sarah Lawrence student from one of my courses to travel with us and help with the field and lab studies. They were bright, interested people, and I felt rewarded by being able to give them the experience. My vetting was not perfect, though. One of them suffered from intense sea sickness and could not help with the diving fieldwork.

The result of all of our work was a detailed description of water movement at the reef surface and a confirmation of my hypothesis about the coexistence of roughheads and spinyheads. Spinyheads experienced faster and more turbulent water flow. Roughheads primarily attacked prey on the surface, but spinyheads rarely did so.[37] Spinyheads had greater success in turbulent flow than roughheads and attacked during a greater proportion of the oscillating flow cycle.[38] Consequently, spinyheads consumed more energy than roughheads, supporting their greater metabolic rates. I never successfully demonstrated that spinyheads win in fights with roughheads, though—a loose end.

The impact of this work is greater than the understanding of two tiny blennies, and the results can be applied to many other species that are not as suitable for study. Understanding the role of water motion in food capture helps with our broader knowledge of the dynamics of coral reefs and gives us some sense of the consequences of reef degradation. Coral death results in reduced physical complexity and lesser turbulence, which affects the food-capturing ability of many fish species, perhaps resulting in their elimination from the ecosystem and, ultimately, their extinction.

Harsh Environments

My father wanted me to be tough, to grow up to be a "Real Man." He hated the gentle host of *Mr. Rogers' Neighborhood*, a children's television show. He saw my children watching it, spit in disgust, and said: "I would never want him in my gang." When I was about ten years old, a tough French boy, maybe a year or two older than me, would chase me every time he saw me. Once, he chased me all the way to our front door, and I tried to get in, but it was locked. I stood there, frightened out of my mind, kicking to keep him away. Eventually, my dad let me in and informed me that next time, he wanted me to "come home with that kid's teeth in my knuckles." I never in my life punched anyone. The only time I ever fought was at age thirteen when I arranged to meet another guy in my class after school to have it out. He was the same kind of wimp I was. I had the upper hand on him, holding him in a lock with my legs when my mum came by, quickly put an end to it, and took me home! I never became a "Real Man."

My father was tall and lanky, with thinning sandy-colored hair and a mouthful of rotting teeth. Except for the sides, the hair eventually fell out, and all the teeth were pulled. When I was little, he played fullback for a Hungarian club soccer team, where he was known as "Knobby." He knew how to kick a ball. A few years later, when he was no longer playing, he took me to a park to kick a ball around. I don't know what came over him, but instead of a pass, he walloped that ball as hard as he could directly at me, hitting me in the stomach and doubling me over.

On another occasion, he took me fishing at Lac des Ormeaux in the Laurentian Mountains north of Montreal. Getting there involved a long, several-mile uphill hike on a very hot day. As he soldiered on, I fell behind, face red as a beet and struggling with a cloud of blackflies surrounding my head. I recall seeing him getting farther and farther ahead and finally disappearing, leaving me alone to follow on the poorly marked trail. In later years, whenever I returned home for a visit, Dad would greet me at the door with a handshake. He never hugged me, even as a child.

I don't know what he was thinking during these incidents. We never spoke about them. I resolved never to treat my own children that way. When Serena was a month or two old, I carried her around in a front-pack. That was in 1978, and seeing men carrying babies was still rare. My father would never have done that, but as I recounted earlier, these harsh tendencies of my dad's didn't preclude his loving me and being nurturing in his own way.

Winters in Montreal

When I was four, five, and six, the arrival of winter was marked by cod liver oil. My mum forced me to swallow a tablespoon of the vile stuff every morning before breakfast. It was a source of the vitamin D I needed for my growing bones. Back then, before global warming, the snow came to Montreal in early November and lasted until April. It was no big deal. Everyone bundled up and did what they needed to get through the season. If a car got caught spinning its wheels, passersby would get behind and push in rhythm, rocking the car until it got traction and left them behind. You had to be careful not to position yourself directly behind a rear wheel, though, because the spinning tires sprayed your pants with dirty, slushy snow. Men used to wear low-cut black rubber slip-on overshoes to give them traction and protect their good shoes from the salty slush. Every bus stop throughout the city had one or two of these lying forlornly on the sidewalk, the result of someone stepping on another's heel as they crowded to board the bus.

It was always cold, and the glass bottles left on the doorstep by the milkman froze. This was before homogenization. The layer of cream that floated at the top hardened into a mound that pushed a few inches above the mouth of each bottle, the cardboard disks that once capped the bottles perched precariously on top. As a treat, my parents melted this cream into their morning coffee. The laundered clothes Mum hung from the clothesline did not dry; they froze. As she reeled them in, she was careful not to snap the arms on the shirts that were as stiff as boards. Those shirts, with their elevated arms, looked like football refs signaling a touchdown.

Giant plows cleared the streets, creating six-foot banks on the sides of the road. Men dug into these to make parking spaces for their cars, and in the process, huge mounds of snow were created. We kids loved playing on these, not only sliding down them but burrowing into them to make cave-like "houses." Mum put me out in a one-piece snowsuit, and I played with my friends for hours, peeing in my pants because I was having too much fun to go in and endure the arduous task of getting my boots, scarf, and mitts removed and the snowsuit unpeeled. Mum would never go through all that in reverse to send me out again.

The snow banks and piles were periodically removed by snow blowers, each with a gigantic maw containing two big steel screws feeding the mounds into a chute that expelled a jet of snow into the back of an adjacent dump truck. The snow blower and truck slowly moved down the street side-by-side, and behind was a string of empty dump trucks creeping along, ready to move up to the blower when the first truck was full and drove off. They seemed like hungry people in a food line, waiting for their turn to be filled. Every once in a while, the papers had a story about some poor little kid who had died trapped in his "igloo" when it collapsed or chopped up by those relentless screws.

When I was five and in kindergarten, my mum would bundle me up in the morning and walk me to Aberdeen School, seven blocks from our house. She picked me up and took me home in the afternoon. For each trip, she also wrapped up my one-year-old sister, Wendy, and put her in a wooden baby sled that had sides and a pull stick. Off she went, towing my sister and holding my hand. One day, on the way home in a heavy snowstorm, Mum trudging along with her head down against the driving snow, we crossed a busy intersection on a green light. The road was covered in hard, compact snow with ridges left by car tires. When we got to the other side, Wendy was not in the sled! We looked back and saw a car run over her little bundled body lying in the intersection. Mum frantically ran back, picked her up, and rushed her to the sidewalk where I was waiting. Cars stopped, people gathered, and I stood beside my distraught Mum, looking up as she clutched little Wendy. She was fine and still sleeping, in fact. But when we got home and my mum undressed her, we saw a clear black tire track across her ginger-colored woolen overcoat. The coat must have fallen open as she lay on the snow-covered pavement. When he got home that night, Dad's anger-tinged anxiety emerged when he confronted Mum: "How could you not know she fell out? Didn't you feel that the sled suddenly became lighter?"

When I was seven years old, living in that rented row house on Laval Avenue, keeping the house warm took some attention. A coal stove in the kitchen was the only source of heat. In cold weather, my father frequently went downstairs to the cellar with its bumpy, packed earthen floor to bring up a bucket of hard, black coal to feed the stove. Periodically, the coal man came in his truck and shoveled coal down a chute, the coal rattling along and filling the air with fine black dust. The burning of coal produced ash and cinders that, if allowed to build up, would smother the fire. To prevent this, the floor of the firebox consisted of three thick steel rods with projecting interlocking teeth. By opening the stove door and rotating the rods with a portable handle, my dad caused the cinders to break up and fall through to collect below. When the level rose enough, he would empty it. The ash contained some smaller, unburned pieces of coal, however, and to recover these, my dad went to the shed out back and sifted the mixture.

The cinder sifter was a heavy metal box twice the size of a shoe box with a steel mesh bottom. It had a two-foot handle that Dad used to shake the box, causing the smaller cinders to fall through, and the precious coal pieces were retained for return to the stove. He kept the device hanging from a nail in the back shed. One summer day, at the age of six, I was messing around in the shed. I don't recall what I was doing, but at one point, I stood below the cinder sifter looking up at it. I might have been trying to take it down for some reason. In any case, it fell down, and the corner hit me on the forehead. Remember, it was a heavy steel box, so it caused some damage, and evidently, some bacteria got under my skin.

In a few days, my forehead was red and swollen. That's when the doctor made a house call with his black bag. In short order, he pulled down my pants and gave me a double-dose injection of penicillin. He said that without it, I would have died. It was 1952, still in the early days of antibiotics, the so-called wonder drugs, and it did its magic. I recovered with no after-effects. Fortunately, I did not develop tetanus (also known as lockjaw). A neighbor's boy my age stepped on a rusty nail, developed tetanus, and was bedridden for months. He lost so much weight he became skeletal and almost died. I can't count the number of times Mum subsequently warned me not to step on a rusty nail, but that did not curtail my rambunctious running everywhere.

The next year, we got a new, improved oil-burning stove—no more coal to schlep from the cellar or cinders to sift (and no more cinder sifter). It was a lot cleaner than the coal but came with a heavy odor. Still the only source of heat in the house, the new stove sat in the kitchen where the old one had been. At its front sat an inverted five-gallon clear glass bottle containing the transparent yellow-brown fuel oil. Occasionally, a bubble would rise through the oil as air entered to replace the fuel slowly flowing to the burner. When the oil level dropped, Dad removed the bottle and took it to the shed to refill from a large tank. Once replaced in its fitting on the front of the furnace, an energetic series of bubbles would rise within the bottle to replace the fuel consumed in its absence.

When the warmth of spring finally came, the street gutters were filled with flowing ice water that sometimes formed great pools where the drains were clogged. This was before street salt was used to prevent snow and ice accumulation. The trampled snow on the sidewalks had been transformed into a layer of clear, hard ice. Many residents, eager for the end of winter, spent their spare time with axes, chopping up the ice and exposing the concrete surfaces that quickly dried in the warm sun. Walking along the sidewalk, I encountered a random sequence of hard ice, wet slush, and dry concrete. Summer was around the corner. There wasn't a prolonged spring with dogwoods, forsythia, azaleas, and other flowering woody plants that we see in New England. Montreal had winter, a day of mud, and then summer.

Our kitchen also had an icebox, which looked like a refrigerator but had no electrical plug. An inside shelf at the top held a block of ice that would slowly melt as it chilled the food. That meant periodically removing the melt water that had collected in a tray. Once a week, the iceman would deliver a new block of ice, which had been cut from a lake in winter and stored in an insulated ice house. Ice delivery was an exciting event for us kids. The iceman would proceed down the street in his horse-drawn cart, stop at each house, lift the thick brown canvas cover, and snag a block with his heavy curved ice tongs to carry inside. A bunch of boys would follow behind the cart with its meltwater dripping off the back, creating a wet streak as it moved down the street. Invariably at some point, the iceman would chip off some ice slivers and throw them to us. What a treat on a hot summer day!

All this was normal to me. We weren't suffering. Recently, I visited the Tenement Museum in New York City. The museum's purpose is to show how difficult were the lives of immigrants to the city in the early twentieth century. On entering, I immediately thought: *This is just like the house I grew up in*. We may not have been jammed in as tightly as the New York immigrants, but the facilities were pretty much the same. I never thought anything of it before. Everyone else living on our street experienced the same circumstances.

Becoming a Father

Watching a Nature program on TV at home, I once saw a video of a giraffe being born. The mother was standing, legs askew, and the baby popped out, falling about six feet with a thump, raising a cloud of dust on the dry ground. It struggled up on shaky legs, and the mom began licking it clean. I thought that was a rough way to enter the world, a far greater shock than the slap doctors used to give babies to start their breathing. Human babies are also a lot tougher than most of us realize, however. They may not thump on the ground with force, but the journey through their mother's birth canal is much more difficult than a giraffe's. Two evolutionary trends have conspired to make this so. Our evolving bipedal locomotion has limited the size of the opening in the pelvis, and the size of the brain has increased greatly. Consequently, humans are born essentially prematurely at the point where the growing brain barely fits through the pelvis. If they waited any longer to emerge, neither mother nor baby would survive.

This difficulty of evolution was made clear to me with the birth of my first child, Serena. The initial hint came from a comment by a nurse in the maternity ward. She told me she could always recognize the babies that were born by Caesarean section because they had round heads as compared with the elongated heads of those who had experienced a vaginal birth. Human babies have a flexible skull to manage the big squeeze as they transit to the outside world.

At the age of thirty-two, after nine years of marriage, Karen and I decided it might be time to have children. We were still ambivalent and didn't want to lose our freedom, but we essentially became less careful with birth control. In short order, Karen was pregnant. We took Lamaze classes to prepare for natural childbirth. She was preparing to manage the pain, and I was preparing to be her coach, but I think my prep was really designed to prevent my freaking out, not that I was going to. As a husband, I tried being as supportive as I could, but as a biologist, I found the whole process fascinating. Karen was sweating and working hard, but her remarkable control of the situation freed me to appreciate the beautifully designed process unfolding before my eyes.

I knew that in babies' skulls, the edges of the bones were not interlocked in order to allow them to slide by each other, and I knew the center of the skull had a fontanel or soft spot, which I had thought was an indication that the bones were not finished growing. But watching little Serena being pushed out of Karen's uterus was a revelation. With each contraction, the top of the skull would push out a little further and withdraw a bit. Also, with each contraction, I saw the brain bulge through the fontanel, and because the skin was transparent, I could even see the brain fissures. The skull was flexible, and the brain was being compressed! The fontanel served as a kind of release valve, letting a little bit of brain escape as the squeezed skull contracted. I suspect this deformation would have broken many connections between brain cells, and maybe it did, but the young pliable brain would be able to establish new ones quickly. Or maybe the cells in that part of the brain did not yet have many connections as an adaptation to the momentary deformation.

In those moments, I came to appreciate what pressures babies' heads are subjected to, and I understood on a deeper level the elongated skulls of vaginally birthed babies. Their skulls are pliable, can be reconfigured by external forces, and, in a couple of days, regain their round baby shape. As I write this, the sonograms of Serena's own developing baby are taped to the refrigerator door, and the little guy has a beautifully rounded head. Little does he know the transforming journey that awaits him.

I tried hard to give Serena a rich environment in which to grow into a strong, independent woman. I also inadvertently influenced my children because I am a biologist. While visiting a friend's house, Charles Darwin's young son, George, asked: "Where does your father do his barnacles?"[39] And I ask the reader: Where does your father do his barnacles? He doesn't do barnacles? Well, biologists' children can't avoid being immersed in their parents' work and have no idea how unusual their experiences are, especially the children of field biologists, because families often travel into the field together. While visiting an old graduate school office mate, Raleigh Robertson, who was then director of the Queen's University field station in Ontario, I was struck by a conversation. As the sun was setting, we walked along a forest path, and against the dark trees, clouds of tiny flies stood out, backlit by the lowering sun's rays piercing through the dense foliage. At that point, Raleigh's seven-year-old daughter piped up: "Look at all the chironomids." Chironomidae is the family name for midges. You never know what will stick.

Halcyon Environments

As I sit at my desk, I can look up at the wall and see two pieces of artwork done by my children when they were seven years old. I keep them to remind me of the very special time I shared with them at that age. One is a papier-mâché fish created by Justin, painted in red, blue, and green with the scales properly arranged. It looks a lot like a parrotfish to me. The other is a painting by Serena of an ocean scene showing views both above and below the water. There's the required sun in the corner, but also a sailboat with Serena leaping off into my arms as I stand on a shark. She also included a flying bird, a mermaid, a living sponge, a sea urchin, little fish, and assorted invertebrates. Clearly, they were taking in the experiences I provided and learning from them, much as I learned from my father by watching him clean fish, manage his aquarium, and grow white worms.

My children's art pieces were inspired by the summers we spent at the West Indies Lab on St. Croix. The whole family took up residence there every June. My two children, from one to thirteen years old, snorkeled, sailed, and were confident in the water and on all kinds of boats. St. Croix became my second home, and I felt at ease there. I knew the people, the special spots, the hidden beaches. I knew where the genip trees were, trees that produced a sweet, tart fruit consisting of a bright green sphere with a thin, hard skin. We held these between our thumbs and forefingers, placed them between our front teeth, and bit down. The skin snapped and fell off as two rigid hemispheres, leaving behind the big, hard spherical seed covered in the most wonderful gelatinous, intensely flavored flesh. I also knew where to find mangos in the forest and enjoyed other local fruits, especially soursop, but the most distinctive St. Croix fruit was the genip.

Every time I emerged from the airplane at the Henry E. Rohlsen Airport, I felt the blast of hot, humid air, air that felt so thick, almost nourishing. It enveloped me like amniotic fluid. It was a return to my origin, the place where I belonged. As soon as we dumped our luggage in our room, I took one of the lab vans and drove the family to a secluded beach. On with the fins and snorkels and into the warm, inviting water—our first immersion of the season. Before us, the awesome display of vibrantly colored fish we were so familiar with. That first snorkel always felt like a homecoming. I was surrounded by fish and corals I knew intimately. Although it had been ten months since I last swam on a reef, it invariably felt like yesterday—no readjustment, no relearning, no lost memory of fish names. I belonged here, and the reef community welcomed me.

Serena and Justin snorkeling at St. Croix, 1989. Photo by author.

The kids were around and in water almost continuously. I was struck and proud when we were on the beach at Sandy Point, watching little Serena, a year-and-a-half old, playing in the shallow water. An unusually large wave rolled in and tossed her about, somersaulting her. I rushed to help her, but when the wave receded, up popped her head with a big smile on her face. A couple of years later, again on St. Croix, I took my students on a scuba diving field trip. We motored out in the Old Horse, a thirty-foot work/dive boat, to a site off Christiansted, where we all descended to explore a degraded reef. Five-year-old Serena happily snorkeled with her mom forty feet above us. When they were older, Justin and Serena would follow our bubbles at the surface while Karen and I worked below.

Not all went smoothly during our wonderful summers, however. One year, we enrolled the kids in sailing classes at the St. Croix Yacht Club across the street from the lab. They had a great time meeting other children and learning the basics of sailing. During one of their practice races, Justin was wearing a trapeze harness attached to a steel cable, and as they were engaging in a complex maneuver, something went terribly wrong. The boat rolled and capsized, and Justin was momentarily sandwiched between the harness and the centerboard trunk. He let out a piercing scream as both were dumped into the water. Serena swam around the boat, expecting to see a limb missing or something equally disastrous, but he appeared to be unscathed. After they righted the boat and made their way back in, however, the water at the bottom began turning red.

The incident happened towards the end of the day, and when Karen and I arrived to pick up the kids, we were told that Justin, then eleven years old, "had an accident." We took him home, but when I saw the extent of the damage, we drove him, whimpering in the back seat, to the medical clinic at the center of the island. I guess an island doctor sees a lot. He was totally unfazed as he stitched up Justin's torn scrotum. During the cleansing part of the procedure, I could actually see his exposed testis—that was interesting. Interesting to me, the biologist, at least. Justin was frightened and mortified. We were sent off with explicit instructions for him to stay out of the water for a week, but four days later, we were on a beach with good surf, and we could not keep him away from it. He cavorted in the waves without tearing any stitches. Word of his injury got around, and each morning, poor, embarrassed Justin had to face a bully at the Yacht Club who greeted him with: "Mornin' Justin, your balls ok?" I might have helped him manage these encounters, but he didn't tell me about them until he was forty years old.

I am proud of my children and proud to have given them unique experiences as part of their development. I love them and would do anything for them, and the scientist in me is still able to look at them objectively. I'm sure that in a real emergency, I would lose that detachment and wouldn't marvel at the biology of the incident, but short of that, yes, it's how my mind works. They are adults now, and I see that they have picked up on this tendency, but I doubt they see it as completely as I do.

A couple of years later, in an attempt to get to know blennies more thoroughly, I kept a dozen in a five-gallon tank on the kitchen counter in our Ossining house. They thrived, spawned successfully, and some lived for four years, an unexpected lifespan for such little fish. But being blennies, they spent all their time in holes in the coral pieces I provided. So, when Serena's friends came to visit, they were always amused that we kept an apparently empty fish tank in our kitchen. "Where are all the fish?"

Whenever I prepared for a research dive, I made sure my accessory, a small net mesh bag, was properly stocked. This "purse" contained: (1) my notepad made of polypaper on which I could write underwater, (2) a wide-mouthed plastic jar with screened cutouts on its ends for keeping blennies alive, (3) five or six syringes loaded with a dilute seawater solution of quinaldine sulfate (an anesthetic), and (4) a couple of test tubes. With this assemblage, I was prepared to take notes and collect live blennies for use in experiments.

On one memorable dive on the east end of St. Croix, I was collecting blennies to bring back to the lab. Justin, a teenager at the time, was my dive partner, and he was collecting blennies, too. We swam out from shore at Lamb Bay, across the shallow lagoon, through a gap in the reef crest, and out onto the exposed forereef where we set to work. As usual, I had my head down, face almost touching the reef surface, oblivious to all but the small patch of reef in my field of view, when I felt a nudge. Turning around, I saw Justin pointing to a dolphin. The animal must have been curious, for after swimming away, it came back to check us out again. We watched it until it disappeared for good, then we returned to our tasks. After coming ashore, we excitedly acknowledged our good fortune to see the dolphin, and I complimented him on his skill in capturing as many blennies as I had. He made some general comment that ended with how easy it was with "sleeping juice."

My father took me fishing and hunting. He showed me how he worked. For example, if you press your knee into the handle of a shovel, it takes much less energy to scoop up a pile of earth. I followed his example with my children in my own way during our summers on St. Croix.

During their college years, both children started elsewhere but completed their undergraduate educations at Sarah Lawrence, where I taught. To my delight, artistically inclined Serena studied a lot of science and went on to medical school. When I was showing slides in my Marine Biology course, one of the students who was in Organic Chemistry with Serena asked if the little girl in the photo was my daughter. It was. But why was I showing a slide of her to my class? The shot was taken on a family camping trip to Acadia National Park in Maine. We had been walking along the rocky shore littered with washed-up seaweeds. I came upon some particularly large, intact kelp fronds and had Serena hold a couple up to provide a size comparison. To this day, she still teases me, saying that I only took pictures of her to provide scale.

Justin surprised me in a different way when he enrolled at Sarah Lawrence. He decided he wanted to register for my Human Evolution course and asked if it would be okay. I checked with Barbara Kaplan, the dean, and she said: "If you two can manage it, the college has no problem." In the first class meeting, as we sat around a table introducing ourselves, Justin started by saying that he was my son. His maturity at that moment, knowing that it was best to make it clear before the rumors started, was striking. After a couple of weeks, he told me I was the same goofy guy in class that I was at home. I took that as a compliment—I always tried to be informal, friendly, and even express some humor as a teacher.

Serena with kelp at Acadia National Park, 1990. Photo by author.

At times, I have gotten carried away while teaching. Perhaps the most conspicuous occurrence happened on a field trip I was leading on a warm summer day. We were standing in shorts beside a lake, discussing I don't remember what, when I noticed some bladderwort in the water. Bladderworts are aquatic carnivorous plants that capture tiny crustaceans. I took off my shoes and stepped into the water to retrieve a sample to show the students. They passed it around and attentively listened while I described the mechanism by which bladderworts capture their prey. The abundant little bladders have a reduced internal pressure, and when a tiny creature touches a trigger hair on one, it rapidly expands, sucking in the prey. But that was not the end of it. I continued my larger discussion for another half hour, standing there up to my thighs in the water with my students lined up on the bank. I suddenly realized how odd that must have seemed to them and casually returned to dry land.

Having Justin as my student presented some awkward moments, though. When I was correcting the first test as objectively as possible, he got the best grade in the class. *Should I be proud or appalled?* I wondered. In my effort to be fair, I thought I had been harder on him than anyone else. Thankfully, although he still did well in the later tests, he no longer had the highest scores, and I felt more relaxed. That was strange—odd to feel better when my kid dropped in the class ranking.

The West Indies Lab

If you drive along the road hugging the north coast of the east end of St. Croix, US Virgin Islands, you will pass over a ridge and descend into a valley. At the bottom lies a cluster of eight moldering buildings amid overgrown plantings of palms, limes, crotons, frangipanis, bougainvillea, and assorted other tropical vegetation. They are the remains of the once vibrant West Indies Lab (WIL).

The West Indies Lab, 1980. Photo by author.

WIL was magical for me, a place where I could interact with renowned scientists professionally and socially. The lab was both a research and a teaching facility. A new class of students from the continental US would arrive each fall and spring semester. During the summer, there would be a couple of courses offered. I taught a few of these courses as part of a barter system. My compensation was full use of all the facilities for the summer: library, boats, labs, and shop, including modest supplies. The whole family moved to St. Croix for those summers and once for a year when I had a sabbatical leave. During the sabbatical year, we rented a condo and purchased a car. Karen took a teaching job, and one-year-old Serena was in daycare. We were thoroughly embedded residents of St. Croix.

Teaching at WIL was so easy. All concepts and descriptions could be illustrated in the real world. Morning classes were supplemented with afternoon snorkeling or scuba diving field trips to the local reefs and other habitats. On some trips, we collected live creatures to bring back to the lab, where there were wet tables—basically large tables with raised edges that held about six inches of water. Fresh seawater was pumped in directly from the ocean, so the creatures remained in healthy condition. Among the things we could do there was demonstrate how corals compete for space via "aggression." We placed corals next to each other, almost touching, and one would inevitably extend digestive filaments over the other and kill the adjoining section. There was a clear hierarchy between species in terms of who would digest whom. But bringing pieces of reef into the lab provides some of the same dangers as swimming on the reef.

One evening, as I was working in the lab, I heard a piercing scream from the area of the wet tables. I ran out and found a student holding her hand and sobbing, her whole body shaking. She had put her hand into the water and brushed against a fireworm, an innocuous-looking six-inch long creature composed of about a hundred segments, each with a pair of paddles topped with white fuzz. She had withdrawn in pain as the worm pushed out thousands of bristles (the fuzz) that were essentially toxic slivers of glass. Although excruciating, there was no lasting damage, and she was fine the next day.

All but one of the students I taught were from mainland USA. They varied in their backgrounds and degree of interest in biology, but none had any prior experience with coral reefs. The exception was a local Cruzan boy who illustrated the dissimilar interests of people with varying backgrounds. He was a curious student, and although he grew up on St. Croix, he did not have a very detailed knowledge of what lives in the sea beyond what the fishermen brought home. Whenever I discussed a new marine creature, he invariably asked: "Is it good to eat?"

I also learned that you cannot always judge the attentiveness of students by their appearance. During one morning class, I was annoyed by a young man who appeared to be nodding off all period. Toward the end of the class, he raised his hand and asked the most insightful question of the day.

We worked hard and partied hard. Jimmy Buffett was our balladeer. The shelves in the basement of the geology lab, "The Cave," were loaded with equipment and literature, all stored in Heineken boxes. You could say I fell in with the wrong crowd, except for all the good stuff. Karen showed more self-control, but I drank more beer in the one year I spent there on sabbatical leave than I had my entire life beforehand. It was there that I occasionally had to put my foot on the floor while in bed just to prevent the room from spinning. John Ogden, the director of the lab, gave me a major compliment when he referred to me as "a happy drunk." I also discovered that the best cure for a hangover is to go diving. Breathing air at high pressure cleared the head. Of course, this is not recommended because safe diving requires a sharp mind. I have to admit that I occasionally behaved irresponsibly that year (recall the nighttime boat excursion to Buck Island), but fortunately, I suffered no serious consequences.

Aside from the faculty on staff, visiting scientists were a constant presence. Mealtimes in the cafeteria were characterized by stimulating conversations about research directions and methodology. We would discuss observations, sometimes made only hours before. For example, John Ogden came in one day after reviewing videotape from a camera he had positioned by a patch of short turtle grass in the lagoon and announced that it was green turtles, not rainbow parrotfish, that were doing the grazing. He said the turtles fed slowly and methodically, carefully inspecting each blade before precisely nipping it off.

The state of the reef environment was a constant topic, and we discussed the possibility of global warming back in 1979. Because I was at a small college with no colleagues in my specific field, these conversations meant so much to me. I felt rewarded and valued for what I knew in a way that only knowledgeable colleagues can provide. That camaraderie was also frequently expressed by serious ribbing. I recall walking alone on the beach early one morning and coming across a dried-out colony of boulder brain coral. I tossed it into the water and was surprised to see it float. I did not know it at the time, but this species has the lowest skeletal density of all corals. A little later at breakfast, I recounted the incident to Bill Gladfelter. He claimed that corals don't float and teased me incessantly about it. Being a world-renowned invertebrate zoologist, he clearly knew this species' dried skeleton could float but never admitted it.

The West Indies Lab, operated by Farleigh Dickenson University, was established in 1972 and quickly developed into a major research location in the West Indies. As time went on, it became more and more valuable as the number of published studies grew, creating a database that gave context for new investigations. More and more researchers saw WIL as a natural location for their fieldwork until everything came crashing down.

Hurricane Hugo, a Category 4 storm with gusts up to 240 miles an hour, hit the lab on Sept 17-18, 1989. Students and faculty huddled in dorms built to withstand storms. The walls were made of solid concrete blocks, and the roof was tied down with internal steel cables that were embedded in the bottom course of blocks. The powerful wind lifted the whole wall an inch or two and dropped it down several times, but the building continued to provide shelter. When they emerged, everyone encountered the equivalent of a bombed-out war zone. A full soda machine was blown 50 yards until it hit a chain link fence. A four-foot piece of plywood was embedded two feet into the end of a cafeteria building beam ten feet above ground. Three-quarters of the houses on the island lost their roofs. There were stories of cows being decapitated by flying pieces of sheet metal roofing material. Every leaf on every bush and tree was gone. All the hummingbirds that had survived the wind subsequently starved to death due to the complete absence of flowers, the source of their only food, nectar.

Betsy Gladfelter, the director of WIL, struggled mightily to get the funding to rebuild the lab, but Fairleigh Dickenson University was in financial difficulty. A major effort to save WIL was initiated. We all wrote letters detailing the scientific value of the lab, and some people went to meetings to plead our case. Nevertheless, the school's board of trustees decided to accept $750,000 of insurance to use at the university rather than $1,500,000 to rebuild WIL. The University of South Carolina had agreed to purchase and run the lab, but this was not to be.

A few years before Hugo, a Moorish-inspired castle emerged on top of the largest hill overlooking the lab. The dwelling belonged to an eccentric Bulgarian, known to everyone as "The Contessa," and she indeed was a Spanish countess. She had signed a contract for first right of refusal to purchase the property and would not release FDU from the agreement to allow rebuilding of the lab.

Sadly, WIL was closed forever.

My Reckoning

WIL was not the only thing to close in my life. As I approached my fifties, I realized I was not immortal. Of course, I knew that intellectually, but up to that point, I lived my life as though I was on an endless journey. My focus was on the here and now. The future would unfold and unfold. No need to think about it. But turning fifty in 1996 forced me to contemplate the forthcoming end of this trip. Not immediately, but not in some unfathomable future either. And if I had limited time, was this how I wanted to spend it?

I loved my kids. Karen and I ran an efficient household, sharing chores. She taught math at a private high school and did much of the cleaning while I bought groceries and did all the cooking and early-morning lunch-making. We never fought. But something wasn't right. I felt an emotional flatness, as though I was made of wood. I craved absorbing conversation, romantic engagement, exciting adventures. I experienced all of that in my marvelous year on St. Croix. Karen was there but marginally involved, not central to all the social excitement. She was a dependable diving partner and research assistant. For that, I am forever grateful.

I thought about starting life anew—the freedom and optimism. On my birthday, Karen, Serena, and Justin sat me down in an easy chair in our living room and proceeded to give me fifty little presents, one at a time. I was overwhelmed. I thanked them enthusiastically and died inside. I was receiving all this love from my children while planning on leaving. They had no idea. I didn't let on. A couple of days later, I took Justin out in the backyard and sat under a tree with him. He knew from my affect that something was up and thought I would castigate him for taking a beer from the fridge. I might as well have hit him on the head with a two-by-four. Thirty years later, I still feel penetrating guilt for all the pain I caused. To Serena. To Justin. To Karen.

I began sleeping on the futon in the basement and then rented an apartment in town. I had a few dates and met a woman I lived with for a year, after which I rented an apartment near school. In September

1999, I placed an ad in the personals section of a new website. Dating apps did not exist yet. A Brigid Moynahan wrote to me; we exchanged phone numbers and had several conversations before meeting for dinner at a restaurant. She was interested in theater and art, which was new to me, and I found her exciting and refreshing. I moved in with her and her two teenage children the following summer. All was good. We did fight occasionally, but I felt alive!

Brigid's daughter, Lucia, was considering colleges in the fall of 2004. She had a strong interest in theater and wanted to attend Sarah Lawrence. The college allowed dependents of faculty to attend tuition-free. For this to happen, Lucia had to be declared a dependent on my income tax for that year. Brigid and I had been living together for five years, so why not? On December 11, 2004, a few weeks before the end of the tax year, we were married, and Lucia spent four tuition-free years at Sarah Lawrence.

Living and Dying

The backyard of our Montreal row house consisted of a small dirt patch, no grass, encompassed by a solid wooden fence. There was a set of wooden stairs leading to the second-floor apartment. It was under these stairs that my dad built a large cage of chicken wire to house the two white bunnies I was given at Easter time. The enclosure contained a little wooden house where they slept, and the floor of the cage was covered with straw. I was seven. It was my job to take care of them.

Every day, I went to the farmer's market one block up on Rachel Street and scavenged the discarded outer leaves from cabbages and assorted other greens to feed my rabbits. The market was a block long with an overhanging metal awning and farmers' trucks, tightly packed, backed up to the sidewalk. I felt small among the hefty, jostling central-European immigrant women in their kerchiefs who pushed their way toward their various destinations. At one spot, women chose chickens from crowded cages. The farmer tied up the selected chickens' feet and threw them down a wooden chute to the basement, where they were slaughtered, gutted, de-feathered, and returned to the buyer still warm. The thought of helplessly sliding down a chute, bouncing and somersaulting with no ability to right yourself, was terrifying. That cruelty bothered me.

I did a good job caring for the bunnies because not only did they survive and grow up, they did what rabbits do and produced a litter. I still vividly remember when the babies first emerged from their wooden house: eight little pure white puff-balls with black eyes and short ears. Like their parents, they thrived, and in time, I had ten adult rabbits. I was kept very busy hauling large quantities of greens home from the market. We did not have the problem of an exponentially increasing rabbit population, though, because circumstances intervened.

The following winter, during a tough financial period at home, I remember standing in the dark shed attached to the kitchen. It was cold in there, and the only light came from a bare low-wattage bulb hanging by a wire from the ceiling. My dad, who grew up on a farm and was not sentimental about animals, picked up a trusting rabbit, held the two hind legs in his right hand, wrapped the thumb and forefinger of his left hand behind its head, and with a quick jerk, broke its neck. And that was that—after a few sporadic spasms, a soft white bunny with dark eyes and a twitching nose became an inanimate bundle of meat. Dad did it again with a second rabbit. He gutted and skinned them, my mum cut them up, and we ate rabbit stew for dinner. I hesitated before that first bite; it was difficult, but I overcame my feelings and ate my pets.

I was too young to know anything about family finances. I don't know if the only way to have meat was to eat these animals. I wasn't traumatized by watching the killings; at least, I don't think I was. It was uncomfortable to witness, but it was quick and probably painless. I was more affected by seeing chickens have their legs tied up and being tossed down a chute. Because my dad hunted and knew a farmer who also hunted and trapped, we also ate deer, bear, and beaver, mostly in stews. Foods that many consider exotic today, we ate as a normal part of our diet, and we were city folk.

Dad had a group of men he went deer hunting with regularly: Bill, Joe, Dez, and Simon. Occasionally, others joined "The Gang," as they called themselves. They would drive up to La Macaza, 120 miles north of Montreal, and stay in a log farmhouse that belonged to another friend, Remy. I went hunting with them a few times but never shot anything except a target. I recall once someone shot a nursing doe, and when we stopped and made a fire for lunch, my dad expressed milk from the dead animal's teat into his hot tea. The animals were gutted in the bush; the entrails were left for the ravens.

Animals and Science

The only way to get at the question of the blennies' diet was to examine the stomach contents, and the only way to do that was to kill the fish. Earlier in my studies, I might have used the phrase "sacrifice the fish," as was the custom, but Rick Miller, my doctoral advisor during my toad phase, rejected that terminology. He said: "You're not doing some ancient religious ceremony. You're KILLING the thing, so say so."

Occam's razor is a working principle in science. It basically states that if several hypotheses explain a phenomenon equally well, the one that makes the fewest assumptions is most likely correct. This approach has been applied to animal behavior in the form of Morgan's Canon: "Do not interpret an animal's behavior in terms of higher mental processes if it can be adequately explained using lower or simpler processes." In the past, the application of Morgan's Canon has led to a view of animals as stimulus-response systems, or automatons, without the ability to suffer. A sudden pulling away, or even a cry, would be interpreted as a reflex, and there was no need to hypothesize a feeling of fear or pain.

This view of animals led to much cruelty in the lab and field. I participated directly and indirectly in some of that. I dissected countless leopard frogs as a student and as an instructor. The standard procedure to prepare a live frog for dissection is pithing, which means inserting a dissecting needle through the back of the skull into the braincase and moving it from side to side to destroy the brain. Although it seems grotesque, when properly done it is very quick, and without a brain, a frog cannot experience anything. The problem is that inexperienced and tentative students can prolong the process, and instead of a quick "lights out," the frog can suffer for minutes.

Although I always found it difficult to end a life like that, I believed that the knowledge gained was adequate justification. I could not justify another behavior of mine during my sophomore year at college, and at the time, I was insensitive enough not even to try. In the 1960s, the teaching labs at McGill University had a frog room attached to them. It was filled with tanks of frogs awaiting their fates in the labs. At the end of lab one day, a couple of friends and I sneaked into that room and stole a frog. We carried it into the arts building, placed it in a water fountain, and retired to a nearby bench to watch as unsuspecting students bent down for a drink and came face-to-face with a live frog. I don't remember precisely, but I can imagine that we giggled at the sudden jumping back of a startled student. There cannot be a more sophomoric behavior than that, nor a more mindless act of animal abuse. But then again, I *was* a sophomore!

In the same year, I took a physiology class. One lab in that course was a demonstration of epinephrine and other chemical mediators of heart rate. The five other members of my lab group and I walked into a small lab and assembled around the central table on which lay a large, anesthetized dog with black fur. It was on its back, legs splayed and tied to the edge of the table, chest cut open, and the exposed heart pumping away. After several applications of different substances by the lab technician and our observations of their effects on heart rate, the dog was given a lethal injection of Nembutal, and it was over. A dog had been killed so that six undergraduates could see the effects of some chemicals on heart rate. I am not completely against dissection or even vivisection, but you sure need a better reason than that.

In my junior year, I enrolled in a biochemistry class. As part of the lab component, my partner and I were given a white rat to care for throughout the term. We fed it, cleaned its cage, and used it in experiments. One experiment required that we sample its blood weekly. The way we did that was to cut a small piece off the tip of its tail and harvest the blood that ran out of the wound before it clotted and stopped flowing. At the end of the term, we needed to kill the animal to investigate the state of some of its organs. The suggested method was to grab the rat by the base of the tail and rapidly swing it through a vertical arc until its head slammed against the lab bench. Once again, another grotesque maneuver, but quick and painless if done right.

Early in my career at Sarah Lawrence, a colleague needed to kill a rat for her class. I don't recall the purpose, but she asked me to do the deed because she could not bring herself to do it. The method of killing had "advanced," and I used a tiny, rat-sized, all-steel guillotine that provided an instant but bloody death.

Mammals aside, we now are learning that even fish have much greater mental powers than we ever imagined. Recent research suggests that they are capable of sophisticated learning on a par with primates.[40] I got an inkling of their mental abilities when I was maintaining a large coral reef tank in the lab at Sarah Lawrence. A striped fish called a sergeant major was acting aggressively toward some others, and I wanted to isolate it for a while. The only suitable container I had was an opaque white plastic jar, so I punched holes in it to allow water exchange and floated it with the fish inside. Every time I checked on the tank, the sergeant major was resting in the jar with one eye against a hole, watching the goings on in the tank. I had not suspected that a fish would exhibit such curiosity. In fact, I believed that they were incapable of that. It changed how I think about those images of thousands of fish suffocating in nets as they are hauled out of the sea. I still eat fish but am not as thoughtless about it. Maybe I'll mature into a vegetarian eventually. I should.

The conduct that is most recent and cannot be excused as from another era involves the blennies I studied. To understand their role in the coral reef community, it was necessary to determine their place in the food web, which could only be done by studying stomach contents. Recall that I placed the fish in formaldehyde as soon as I surfaced from a dive so that digestion was stopped and the stomach contents preserved. Initially, I put a squirt of anesthetic into the bag of water containing the fish but found that this made them regurgitate some of their gut contents. The only way I could get a representative sample of what was in their guts was to put the formaldehyde directly into their water. When I did this, the fish swam about frantically and eventually died. I no longer could attribute their writhing to simple reflexive stimulus-response. They may have been in intense agony. I hated to do that, but the only alternative was to leave blank an important piece of the picture. That was my decision. I didn't like it, but I did it.

I love Nature. I marvel at the beauty of all living beings, yet I do these awful things. I minimize the harm, but I want to gather the information to help me understand. This understanding can help us value a delicate ecosystem and give us motivation to preserve it. The scientific establishment now takes a very different view of animals. Since 1986, the National Science Foundation has required universities to have

Animal Care and Use Committees that ensure the humane treatment of animal subjects. We cannot eliminate all animal suffering in the name of research, but we can work as hard as possible to minimize it.

Animals have always caused suffering in each other. Early one fall, I found a tomato hornworm in my garden. It was a big, fat, bright green caterpillar with a little tail poking out the rear end. I wanted to see the moth it would become, so I put it in a jar and waited for it to pupate. In short order, it transformed into a shiny chestnut brown chrysalis.

To simulate winter and stimulate dormancy, I placed the jar in our kitchen fridge. (Be careful when you open a biologist's fridge; you may be surprised at what you find. My daughter's friend got a shock when she opened our freezer to get ice cream and encountered a bat in a plastic bag.) After six weeks, I took the jar out of the fridge and placed it on the kitchen counter to complete its development into an adult moth. In time, the skin of the chrysalis split, and out came a . . . FLY!

Talk about a surprise. Instead of a beautiful five-inch hawk moth, I was presented with a hairy fly, not unlike a very large housefly. I dissected open the "empty" chrysalis, and inside I found a large, brown, ovoid structure—the pupal case of the fly. The hornworm I had collected in the fall contained the fly larva, which first ate the non-essential tissue of the hornworm and then finished the job in the chrysalis before pupating inside. With a little research I learned that this was a tachinid fly, a large family that are parasitoids of many insects.

Several parasitoids afflict tomato hornworms. Parasitoids, unlike parasites, always kill their hosts. The most common one is a little wasp that lays fifty to seventy eggs under the skin of a small tomato hornworm. The eggs hatch into larvae that eat the hornworm's internal structures but don't immediately kill it; in fact, it continues to grow. Along with the eggs, the mother wasp injects a virus that suppresses the immune system of the hornworm so it cannot reject her eggs and larvae. The virus also alters the hornworm's development so that it grows larger than normal and does not transform into a pupa. Eventually, the larvae bore out through the hornworm's skin and pupate, forming a halo of small white ovoid cocoons on the caterpillar's surface. At this point, the hornworm dies.

A tomato hornworm. Photo by author

So, if you see a big, juicy tomato hornworm in your garden, there is a chance it's a vehicle containing a host of wasp larvae or a single fly larva selectively consuming its insides. In the 1979 movie *Alien*, a creature develops inside a man and emerges from his chest in a frightening, gory scene rated #2 in BRAVO's "100 Scariest Movie Moments."[41] Science fiction is science reality to the poor tomato hornworm. A common saying among scientists impressed with their findings is: "You can't make this up."

I have no idea if tomato hornworms experience pain or fear. Maybe they are the automatons we used to think all animals were. But I'm pretty sure an antelope can experience pain and fear. Maybe not exactly as we do, but in a way that means they can suffer. It may not be a big deal for a tomato hornworm to be eaten alive from the inside out. But it is a big deal when an antelope is having its hindquarters eaten by a lion while it is still alive, or a deer is being suffocated to death by the constrictions of a python, or a chimpanzee is being beaten and bitten to death by a gang of chimps from a rival troop. Pain, fear, and suffering occur throughout the natural world. Nature does not have a conscience. It is value-free. It just is.

So, what do we make of our treatment of animals? Are we part of Nature, and we simply do what we do? Or are we separate? Do we have a moral imperative not to cause suffering in animals? At what point between becoming bipedal four-and-a-half million years ago and the industrial revolution two hundred years ago did this happen? I am not a philosopher and will not try to answer these questions. I can't. But I have empathy and feel terrible when I see suffering, be it human or animal. To avoid these awful emotions, I minimize my role in animal suffering.

Octopus Stories

One evening in December 1969, during my first semi-tropical experience in the Florida Keys, we walked along a canal, peering into the water with the help of a flashlight to see what nocturnal creatures were about. We spotted an octopus, about a foot-and-a-half from arm tip to arm tip. It was crawling along the base of the concrete wall. I wanted to catch it for a closer look but was aware that they can bite with their parrot-like beaks. I overcame my hesitancy and quickly scooped it up into a bucket of seawater with my bare hand. We brought it into the motel room to get a better look, and it promptly crawled out of the bucket onto the floor, its eight arms suckered to the tiles. Trying to get it back into the bucket, I pried two arms off the floor, but as soon as I released one arm to reach for another, it would snap back onto the floor. I needed eight hands, one for each arm, but how do you get four people huddled close enough around an octopus to lift all eight arms off the floor? I don't recall exactly how we did it, but we managed and returned the creature to the same spot in the canal where we found it.

I have been impressed by marine creatures in unexpected ways, ways that blunted my scientific skepticism. When someone tells me about an unusual observation, Brigid says I knit my brow and get this look that indicates: *I don't believe that.* I guess I should be more vocal about what I'm thinking. She says that if I didn't experience it myself, I don't accept it. I do agree with that characterization. I once observed octopus behavior that I would not have believed had I not seen it with my own eyes.

I was on Takapoto Atoll in French Polynesia on a field trip associated with the Fifth International Coral Reef Congress in 1985. Takapoto is a remote atoll with a small population. When our group of twelve scientists arrived by airplane, we were escorted to a meeting house where we were introduced to the village elders, and we each told them about our interests in coral reefs. We spoke in English, and our guide translated our words into French, after which a member of the community translated the French into Tuamotuan. I have no idea what came out of the other end of that language tube, but everyone smiled, nodded, and seemed satisfied. After some refreshing fruit drinks, we were shown to our quarters—an assemblage of tiny lime green plywood huts roasting in a shadeless patch of tropical sun. The huts were so tiny that we called them Smurf Village, after the popular TV cartoon characters of the time. Following lunch and an afternoon being shown how they culture black pearls in their lagoon, it was time for a wonderful feast of fish, pig, and tropical fruits.

In the evening, the huts in Smurf Village cooled down enough to be bearable, and we eagerly settled down to a solid night's sleep. As in most tropical houses, there was no glass in the windows; they were either open to let the breeze in or shut with wooden panels against rain and storms. We drifted off to sleep, relishing the breeze and soothed by the distant roar of the surf on the reef—until a series of loud shrieks nearby jolted us awake. Scrambling out the door to see what was happening, we were met by a quickly assembling mass of sleepy scientists. As word spread around, we discovered the following: as is the custom in sexually open Polynesia, a boy had come to the open window of a young woman scientist in our group, and she awoke with a terrible fright. The unexpected response sent the boy running off in a terrible fright of his own. No lasting harm done; only elevated heart rates and a cultural misunderstanding that was quickly cleared up. Takapoto may have been remote, but it was not lost in time. The communal building at the center of the island had a generator that powered a dozen or so machines playing video games.

But back to octopuses. The next day, I was walking with several biologists and geologists on a flat rubble beach exposed by low tide. Such beaches consist of small broken pieces of coral, white and eroded. They dry out quickly when exposed to the air. We came upon a tiny shallow inlet about four feet wide, twelve feet long, and a foot deep. I looked in to see what life it might hold and saw two octopuses. The larger one was approaching the smaller one, which was slowly retreating in seeming fright until it released a cloud of black ink and jetted towards the blind end of the inlet. But the larger one kept advancing, and that's when the unimaginable happened. The small octopus left the water and, with flailing arms, crawled across the rubble, dry coral chunks clinking and rolling into the water as it shuffled along. The little octopus moved parallel to the water's edge until it was past the large one, still in the water. At that point, it reentered the water and jetted away toward the open end and freedom. Just think what a foreign, inhospitable world dry land is to a soft-bodied creature with neither waterproofing nor lungs. Breaking through the top of its world, navigating there, and reentering the water where appropriate was both a physiological stress and a remarkable conceptual leap I never expected.

I regularly gave this account to my marine biology classes and backed it up with another anecdote I had heard. The story is that in a lab somewhere, investigators would find crabs missing in the morning. In an effort to determine the cause, they set up a video camera at night. The tape showed that an octopus was leaving its tank, crawling over the tabletop to the crab tank, catching a crab, and eating it before returning to its home tank. Recently, in an effort to get more detail, I searched the internet and found several versions of this story. I also came across a page that investigated these accounts and concluded that it was an urban myth—no credible documentation of the accounts could be found. How disconcerting to know that for many decades, I had been presenting an urban myth to my students as fact!

There have been a few impressive accounts of octopus intelligence in the peer-reviewed scientific literature, however. In 1992, Graziano Fiorito and Pietro Scotto trained octopuses to attack either a red ball or a white ball by rewarding a correct choice with food and an incorrect choice with a mild electrical shock.[42] Once these animals were trained, they were used as demonstrators, performing the choice while other naïve animals in an isolated tank could observe. These test observers were then presented with the red and white ball choice and, without any reinforcement, chose the correct ball for their trials. This was the first demonstration that octopuses could engage in observational learning.

Other documented feats involve screwing the lids off jars to access food or to escape. Think about it; the octopus turns the lid, and nothing happens. It turns the lid again, and still nothing happens. Finally, after several turns, the lid comes off. That behavior represents remarkable foresight and persistence. There have also been a couple of other documented cases of octopuses leaving the water, including a spectacular video of one vaulting out of a pool to capture a crab walking above the water. The octopus dragged the crab back under the water and into its lair beneath a rock.

Octopus intelligence is unexpected in the context of their life cycle. Most creatures that exhibit such impressive flexible behaviors are social, long-lived, and reproduce over an extended time frame. Octopuses are solitary, generally live one to two years, reproduce once, and die. This life history presents fewer opportunities for learning and no capacity for passing on knowledge to offspring. Such animals are generally highly pre-programmed in their behavior. Yet octopuses are bright, flexible, and capable of surprisingly sophisticated behavior. Recognizing this, the European Union gives them the same protections as vertebrates in relation to research practices. Of course, this "honorary vertebrate" status can only be conferred by us vertebrates.

A Soft Spot

As an ecologist, I've been trained to think primarily about the health of populations and ecosystems. The subject is the dynamics of biological communities, and individuals are simply building blocks for these systems rather than subjects in themselves. When I think of the differences between me and my students, one example from the news comes to mind. In the fall of 1988, three gray whales were trapped by growing sea ice in the Beaufort Sea off Port Barrow, Alaska.[43] Gray whales were not endangered. In fact, their numbers were estimated to be larger than in pre-whaling days. To free them, a great assortment of individuals and agencies came together, including two Soviet icebreakers, the *Vladimir Arseniev* and the *Admiral Makarov*. News agencies all over the world breathlessly followed the rescue effort daily. One whale died, and the fate of the other two is unknown. The operation cost in excess of one million dollars. I recall thinking then about how much more good could have been accomplished had that money gone to so many other environmental causes. Although I am not inured to the suffering of individuals, their costly rescue is not a priority given our limited financial resources and the natural processes of the living world.

On a bright spring day in 1976, however, as I sat in my office at Sarah Lawrence talking with an undergraduate student, another came to the door, gingerly carrying something in a box. She handed it to me. I lifted the top and saw a tiny gray squirrel; it was skinny and bedraggled, with some blood around its nose. The student said it had fallen from the roof of Westlands, the administration building.

"Can you take care of it?" she asked.

Not wanting this extra responsibility, I said, "You know, wild animals die all the time. It's not our place to interfere." She wouldn't buy that, which is a common attitude among people not steeped in ecological and evolutionary science. Once, a student on a marine biology field trip even asked me if we could save a starfish that a seagull was eating.

"But what am I going to do with this little guy?" she persisted.

"It probably won't survive even if we try to save it."

"But you have to try."

With resignation, I said, "Okay, I'll see what I can do." Like most people, I have a soft spot for warm, furry things.

I took the squirrel home and started feeding him milk with an eyedropper. Eating eagerly, he improved but had to be fed many times a day, so I carried him about in the breast pocket of my shirt, where he would curl up and sleep between feedings. He traveled like that as I rode my bike to school, and all over campus, to the post office, the cafeteria, and the library. My students began calling him Eugene. One time, as I sat behind my desk conferring with a student about his project, Eugene turned over in my pocket, and his tail flipped out. I have never seen a more surprised look on anyone's face!

The little squirrel grew, began eating solid food, and had free reign of our third-floor apartment in Yonkers. Several books on the shelf had their corners chewed off, and we were constantly wiping up droppings. He became quite territorial about our space, and when Floyd, my old friend from graduate school, came to visit, the squirrel leapt up and bit him on the hand, drawing blood. This was the same little guy who would comfortably sit on my shoulder and climb all over my head. Clearly, it was time for our arrangement to come to an end, even though our apartment would feel empty without our fluffy squirrel scampering about and crunching nuts.

We'd lived with him for five months, and by then it was fall. The oak trees were dropping a heavy crop of acorns. On a warm, sunny morning, I took the now-grown squirrel to a quiet area of campus and released him. He readily climbed into a tree, and I headed to my office, feeling as beneficent as the student who'd brought him to me in the first place. This happened before the need for a long education period to prepare captive-raised animals for life in the wild was reported in many publications. Did he find food? I'm quite sure he did. Did he find shelter? Perhaps. Did dominant resident squirrels bully him? I have no idea. Like the gray whales in the Beaufort Sea, he went off to an uncertain fate.

Cats

On a beautiful, sunny, early fall day in 2008, Brigid and I went to an outdoor farmer's market in Montclair, New Jersey. We talked to some farmers, bought some vegetables, headed for the exit, and there on our right, the last stand before the street, was a booth operated by PAWS, the Pound Animal Welfare Society of Montclair. They had a few cages with the cutest puppies and kittens you can imagine. It was a bit like the candy shelves at supermarket checkout counters—an irresistible enticement to encourage an impulse purchase on the way out. Not that PAWS was irresponsible. They went to great lengths to ensure that the animals they released were well cared for, including having adopters sign a contract stipulating the conditions for proper care.

For some time, Brigid had been saying we should get a cat, but we didn't do anything about it other than my saying: "If we get a cat, I'll be the one to clean out the litter box." We lingered at the booth, checking out the kittens. Observing our interest, the attendant came over to talk. She said: "We have many more cats at our facility just half a block away on North Willow Street." We were having a relaxing Saturday, so we decided to check the place out.

The unit had cage after cage stacked up four high, running down the length of the narrow building. We walked along looking at cats and kittens, each alone in a cage. Then we came across two half-grown cats together, one lying on top of the other, the bottom one purring loudly. They seemed relaxed and so attached. At that point, we decided that two cats would have a much richer life than one alone. And that was that.

Being rescue cats, they did not have "papers," but we soon discovered that, although the veterinarian's office had labeled them "domestic short hair," they were Russian Blues. They had short, thick gray fur, beautiful green eyes, and gray noses instead of the pink of most cats. They were frightened when we first released them in our kitchen, so I built a little shelf under an open counter they could shelter in, and they hid there for a couple of days. At one point, as they became more relaxed with us, one raised its tail straight as a rod into the air. I exalted—that tail rise is a cat greeting sign. We finally had our connection. They had accepted us as part of their social unit.

Our cats, major predators. Photo by author.

Both the people at PAWS and the vet told us we should keep them indoors. They informed us that outdoor cats contract diseases and have accidents; they have a shorter lifespan than indoor cats. But we had a lovely big yard with a lawn, bushes, trees, and wildlife. It seemed that outside was a much richer environment for our cats. We let them out during the day, and they wandered extensively. Once, I saw them standing at the peak of the gabled roof on our neighbor's garage. We always called them in for their evening feeding and kept them in the house for the night.

Then they did what cats do. They brought home gifts for us—birds, mice, shrews, chipmunks. We would find these dead animals in various places in the house. But sometimes, they would bring in live ones, and we would chase the terrified animals under couches and behind radiators until we captured them and released them back into the yard. But our cats, being Russian Blues, were larger than your average house cat and capable of more. One day, I went into the basement and found a dead rabbit oozing blood on the concrete floor. This carnage was getting serious.

Montclair had a deer problem at the time. The wild eastern whitetail deer had become quite urbanized. I recall taking a walk in the neighborhood and seeing a woman shouting in her yard, chasing three deer away, and thinking: *Huh, why would she do that?* Evidently, the deer had become pests—chomping away at the expensive shrubs suburbanites plant around their houses. In time, I experienced the same thing. It got to the point where I actually picked up small stones and threw them at the deer I was chasing in our yard. Sometimes, there would be as many as five at once.

One morning in early summer, while I was quietly reading on the back deck, I looked up. About ten yards away, across the lawn in the shrubbery, was a doe and two little spotted fawns with shaky matchstick legs. They were quite relaxed, and so was I. I certainly wasn't going to chase these little guys away, so I just continued reading, looking up occasionally to take in the spectacle of newborns with their mom. They hung around, and I eventually went about my day. In the afternoon, I went into the basement and found a dead fawn! I had never heard of cats killing anything that size. Although much taller than our cats, the fawns probably weighed about the same, or even a bit less, than our fourteen-pound cats. I have no idea how the deed was done. I am unaware of domestic cats hunting cooperatively. My guess is that one of them singlehandedly took down this little fawn and dragged it down the steps into the basement.

Our cats now spend half the year as apartment cats in New York City and the other half as indoor cats in beautiful rural Truro, Cape Cod. We could let them out on our Truro property; there is a rich natural environment for them to explore. But that natural environment harbors coyotes that are known to kill cats and small dogs. Additionally,

I like to keep our property as a mini nature preserve and do not want the carnage that domestic cats bring. Cats are a major cause of death of birds and small mammals, and our cats have demonstrated they are no exception. So now, they sit by the patio doors watching the birds and red squirrels go by, sometimes sitting side-by-side following the animals with their heads swinging in unison like spectators at a tennis match. Their tails twitch with excitement, just like they do when stalking prey. It must be frustrating for them to "look but don't touch." Well, we all have to make our compromises. And for all these years, I have been the one to clean out the litter box.

Discovery of a Species

I was on another research dive at Carrie Bow Caye in Belize. It was 2001, and I was working with Jim Tyler, a colleague at the Smithsonian Institution's National Museum of Natural History. Part of an experiment we were conducting involved removing the blennies from the upper parts of branching elkhorn coral to see if the fish in the lower portion would move up to occupy the vacated cavities.[44] I was inspecting the underside of a large, flat branch, looking for blennies to remove. The section of coral was dead and coated with coralline algae. Coralline algae tightly cover hard surfaces with sheets of living tissue embedded in stony calcium carbonate. It looks like a thick layer of pink paint. And there, sticking out of a hole in the coralline algae, was a pink head with prominent eyes. It was the same size and was positioned in the hole the same way as a blenny. But it was not a blenny. "My" blennies don't come in pink, and their heads are shaped differently. Wanting to know what the mystery fish was, I captured it and brought it back to the lab.

Back on shore, using a microscope, I inspected the fish carefully and used several published guides to determine that it was a clingfish. Clingfishes are characterized by their modified ventral fins that form a sucker allowing them to grip tightly to surfaces in the face of strong currents. Although I knew it was a clingfish, I could not match it with any of the described species. I showed it to Jim, who, as a specialist in fish systematics, was much better equipped than I to assess the taxonomic fit. He agreed that the specimen was not described in any of the literature available to us. He said: "I'm working on a project on clingfishes with Jeff Williams. Can I take it back to Washington?" I agreed, placed the fish in formaldehyde for a day, then stored it in alcohol. This is the standard way to preserve fish specimens.

Over the next couple of years, Jim and Jeff spent countless hours examining the hundreds of specimens of clingfishes from collections in their museum and institutions around the world. They were performing the tedious work of confirming existing species and describing new ones. They were glued to their microscopes, counting things like "pectoral-fin rays . . . segmented caudal-fin rays . . . precaudal vertebrae" and a host of other characters. After careful analysis, they recognized ten species, five of which were new. Each new species was described in exquisite detail in a unique way that would allow others to match their specimens to identify them.

In describing five new species, Jim and Jeff had the opportunity to name them. All scientifically described species have unique Latinized names, which is essential because many individual species have several common names, and that leads to confusion. For example, I once witnessed a commotion in a restaurant on St. Croix because a patron was upset that dolphin was being served. The item on the menu was dolphin, the fish, also known as mahi mahi or dorado, not dolphin, the mammal that at the time starred as "Flipper" in a TV series. Although we scientists are comfortable with unambiguous Latinized names, I'm not suggesting they be used on menus.

Latinized scientific names usually indicate a description of the species' characteristics, the place where it was found, or the name of a person the author wants to acknowledge. But scientists can be playful, even frivolous at times, especially if they are naming many new species. For example, Terry Irwin of the Smithsonian Institution named 113 new species of the beetle genus *Agra*. I don't know if he ran out of ideas or simply needed a break, but now the official permanent names of four of the species are *Agra cadabra*, *Agra phobia*, *Agra vate*, and *Agra vation*. Maybe the many long hours staring into a microscope and recording minuscule structures drives some people batty. A little fun takes nothing away from the serious hard work.

However, there is an infrequent, sinister side to scientific names. In 1937, the Austrian entomologist Oskar Scheibel, discovered a brown, eyeless beetle living in Slovenian caves. He named it *Anophthalamus hitleri,* after one of his greatest heroes.[45] Horribly and ironically, the species is now endangered because of demand by Nazi memorabilia collectors. As of this writing, there is a movement to change the species name, but there is some resistance for several reasons, not least the importance of stability of the nomenclature.[46]

One day, out of the blue, I received a phone call from Jim. He said they were writing the paper and had a question for me: "Do you want to be a coauthor, or do you want the species named after you?" It didn't take me long to make the decision. I felt my input to the work was simply to provide a specimen; I didn't have their expertise and hadn't done the hard lab work that Jim and Jeff had done. I didn't want to be a coauthor out of fairness, but I also had a selfish reason for choosing to have the species named after me. It is a form of immortality in the sense that this fish species will bear my name forever. So, if you ever look up the technical literature on clingfishes, you may run across a species called *Tomicodon clarkei*.[47] It has also been given the common name "coral clingfish."

Since then, I came across the following quote by Peter Wohllben in his book *The Secret Wisdom of Nature*:[48] "Discovering a species exists doesn't say much except that it has been spotted somewhere at least once and we have a description on file." To my knowledge, nobody in the world other than me has seen a live coral clingfish. So, should I feel blessed to have had the privilege of seeing such a rare, or at least very cryptic, species, or is this observation a pretty meaningless piece of information? Maybe I've achieved worthless immortality. Only time will tell, and I will never know.

Communicating Science

I had little time for data analysis in the field; all my time was spent preparing equipment, attending to diving gear, evaluating study sites, and assorted other chores in need of immediate attention. Once back at the office, my analysis could begin. One goal was to search for patterns that would help me understand the functioning of the system. Another was to test hypotheses I had formed before doing the fieldwork. This work involved a lot of numerical summarizing of data in many different ways and the application of a great variety of statistical procedures. Although the data are sometimes presented in tabular form, graphs are used whenever possible.

"Clarke would rather use a graph when a simple paragraph would suffice." This quote from a student evaluation of one of my courses early in my career says so much about how people take in information. It always elicited a laugh when I told it to my science colleagues at other institutions because it was so nonsensical to them—we all shared a visual view of the world. But it wasn't nonsensical to that student. His mind was more tuned to the word than the picture. I learned from him that I, as a teacher, should understand this difference and account for it in my teaching. But the truth is that if you try to create a verbal or written description of the complex relationship of two or more variables, it's very difficult to do clearly. So, I kept using graphs because they are by far the clearest and most efficient way to express quantitative relationships. But I also went to great pains to define the axes, units of measurement, and the meaning of the shapes of the curves on the graph. I helped, and hopefully taught, my students to read the graphs.

It is very easy to create a graph. Many computer programs will do that for you. All you need to do is input a list of paired numbers (for example, the number of eggs laid and body size), select a graph type, and out comes a finished product with labels on the axes. Then, you can modify many features of the graph to maximize the clarity of the presentation or highlight specific aspects of the relationship. There is also an esthetic to graphing—some versions are simply more pleasing to the eye. This may not make them more effective at conveying the information, but I often put in extra time to get them just right. That meant experimenting with font size, line thickness, symbol types, sample sizes, statistical significances, and associated data. A key to effective graphing is keeping it simple, however. If you try to convey too much information in a graph, it becomes confusing. Just because a graphing program can express a two-dimensional relationship with three-dimensional columns doesn't mean you should. I have often seen published graphs that are unnecessarily complicated, less pleasing to my eye, and not "clean" in the sense of being as simple as possible and thus easiest to absorb.

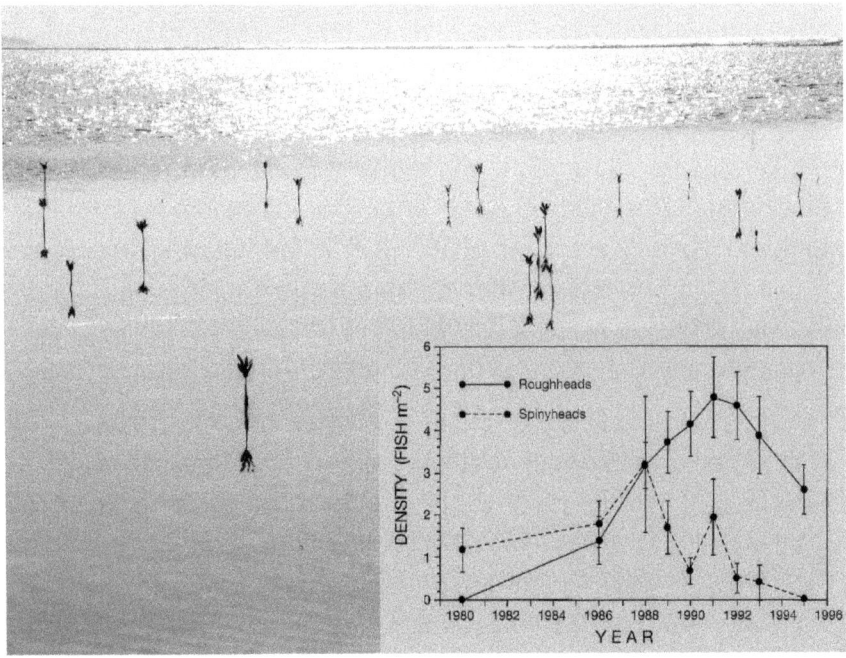

Error bars on a graph and "error bars" created by reflections of mangrove seedlings in quiet, shallow water at Middle, Cay, Glover's Reef, Belize. Photo by author.

In 1969, when I was a graduate student, graphing was a more arduous task than it is now, and the frivolities tempted by modern graphing were not feasible. I would start with a piece of graph paper with a specific grid pattern. There were many kinds available with different scales, such as arithmetic, logarithmic, and semi-logarithmic (the last two with differing numbers of cycles). Having assembled the coordinates to plot, I carefully moved my pencil along one axis and then the other and marked a point at the intersection. I graphed all my data points in this way. Then, I would place the fitted line on the graph. To do this I performed a standard statistical calculation, called the "method of least squares," that resulted in an equation describing the relationship of the two variables. Using this equation, I calculated the coordinates for two distant points on the graph and plotted these. Joining them with a straight line gave me the "best fit" to the data points.

If the data were to be represented by a curve, I would draw it using a set of French curves. These are clear plastic shapes with the edges outlining curves of varying radii. Segments could be combined to generate complex curves. Another device was a flexible plastic curve template designed to bend smoothly. Either of these could be traced with a pencil to fit a curve through the data points. With a straight line or a curve drawn in, my graph was precise but rudimentary, not the clean, black-and-white version with labels required for publication.

To turn this into a sharp graph for a scientific journal, I taped a piece of onionskin paper on top of the graph paper. Onionskin is white but translucent, and the rough graph showed through the paper. I then traced all the lines with black India ink and added the data points, numbers on the axes, and labels by using transfer lettering, which came on transparent plastic sheets containing letters, numbers, or symbols. By carefully positioning the character on the graph and rubbing the sheet with a dull pencil, I transferred the character to the graph and carefully peeled the sheet off. When all the transfers were complete, the graph was done. I taped the onionskin onto a piece of white typing paper, and the black lines and lettering now showed up in stark contrast to the pure white background. This completed graph represented much work and was too valuable to lose. When submitting a manuscript to a journal, I had the graphs photographed, and the photos were submitted.

Today, rather than being arduous, graphing is a pleasure. I can change the scales, try different symbols, make any number of alterations, and immediately see the result on my screen. When satisfied, I can hit "Print," and out comes a clean, publishable graph. But there is still something to say for graphing by hand. I believe that students get a much deeper appreciation for a relationship when they have to choose the scale and plot the graph data point by data point on a piece of paper. By creating graphs manually they are better prepared to fully understand the graphs they see in publications (or on the whiteboard).

In a scientific report, all the analyzed data appear in the "Results" section. The rest of the paper consists of a review of previous work on the subject, a presentation of the methods used in data collection and analysis, and a final "Discussion" section in which the findings are placed in the context of theoretical considerations. The process of building my detailed reports resulted in distilling a portion of the complex, dynamic reef of multitudinous fish, diverse invertebrates, algae, and plankton into a neat presentation in a scientific journal. Every scientist experiences considerable satisfaction in seeing their work appear in print. This sharing of findings is essential to the scientific process. In this way, I felt part of a great social network of collaborating investigators demystifying the workings of the most complex ecosystem on Earth.

Size Does Matter

Brigid likes to tell the story about how I study tiny fish, but for our first date, I bought tickets for Laurie Anderson's techno opera *Songs and Stories from Moby Dick*. I think her amusement stems from a sense that I needed to compensate for the dinkyness of my study subjects. There's no question that scale has a large impact on how we react emotionally to encounters. A whale evokes awe, a sense of something magnificent, greater than you, moving in a deliberate manner, magnificent. A blenny evokes no such reaction—maybe just an emotionless curiosity at this tiny, speedy, energetic being bouncing among the corals. I have often wondered about the effect of size on our reactions. For example, an injured whale or elephant evokes much more empathy from most of us than does an injured mouse or grasshopper. Why? What evidence do we have that large things suffer more than small things? What evidence do we have that large things are like us, only more so, and that small things might be like us but less so? Maybe it's because small things are more numerous, and by some weird mental twist, we feel that the capacity to suffer is diluted among so many more bodies, reducing the ability of each to experience anguish.

I shared Brigid's amusement about my apparent need to compensate for the tiny fish I studied. I occasionally felt the need to back off my close-up face-against-the-reef view of the world. While working, I would stop and gaze every time a shark, a dolphin, or a hawksbill turtle passed by. While these were serendipitous encounters, I once actively planned a visit with representatives of the megafauna. It was in Hawaii in 2003. Sarah Lawrence College had flown me out to give seminars to alumni in the Faculty on the Road program. That commitment was for a week on Oahu, and I thought that it was a wonderful opportunity not to be passed up, so I spent an extra week exploring Kauai and Hawaii, the Big Island. On the Kona coast of Hawaii, I found out about a dive shop that took people out at night to encounter manta rays, which are immense, flat fish with wing-like extensions.

The largest species has a wingspan of twenty-three feet. They feed on plankton and are totally harmless to humans.

Unlike skates, which have a similar shape but live on the bottom, rays fly through the open water. Their wings beat slowly, and their movement is very graceful. I observed the most striking example of this elegance in 1973 off of Turtle Rocks, south of Bimini in the Bahamas. As I worked in twenty feet of water, four spotted eagle rays swam into view. Eagle rays are smaller than mantas, with wingspans up to ten feet, but these were about six feet wide. Unlike mantas, which are an indiscriminate uniform gray color, spotted eagle rays have a rich chestnut brown back dotted with hundreds of two-to-four-inch white spots. Also, unlike the short tails of mantas, spotted eagle rays have long whip-like tails, sometimes up to twice their body length. I don't know what prompted this, but all four began to perform summersaults. As they curled, their milk-white bellies were exposed, and their tails, trailing in the loop, actually touched their snouts so that each ray formed a vertical circle, like a hoop. They slowly executed this fluid maneuver several times and swam off, leaving me wondering what that was all about.

As the sun was setting in Kona, twelve of us piled into two boats and headed out, paralleling the shore and anchoring off a point in about thirty feet of water. We were paired up into diving buddies, and each person was given a strong flashlight. We were instructed to kneel on the sandy bottom and direct our lights upwards. When you shine a light in the sea at night, you get the same result as turning on a light in the woods at night—a great assemblage of tiny critters is drawn to the beam. Instead of moths and a few other insects, the ocean cloud is composed of swimming worms, shrimp, and a wide variety of lesser-known invertebrates. The dive shop took guests to this spot regularly, and the mantas were conditioned to expect food. I settled in my spot, turned on my light, and immediately saw that my beam was much weaker than the others. I saw mantas looping down for plankton in the bright beams of the other divers and sat there feeling alone and somehow mistreated. And then, out of the darkness, a gigantic shape descended towards me. I saw an open mouth, four feet wide with soft white skin lining the inside, and the five paired gill slits, dark channels on either side through which water flowed. The mouth came straight for my head and swerved upward just before reaching me. I didn't move. The immense ray passed a foot or two over my head, and I was buffeted by the strong currents generated by that massive body driving through the water. The powerful flow alerted me to the strength of those slowly beating wings. Wow! That image is imprinted in my brain like few other memories.

For years after that encounter, I continued my study of inch-long blennies. No matter how much I learned about them and how much I cared about the health of their populations, no blenny could ever come close to creating the emotional impact on me as did that gargantuan manta ray.

Science is Conservative
(But Not How You Might Think)

To the surprise of many, this beautiful, creative process is conservative. I don't mean politically conservative. I mean that for science to be an effective vehicle to uncover reality, its practitioners must not wander beyond what is empirically demonstrable. Results don't appear out of nowhere like magic. If you ask: "What color is that barn across the meadow?" A cautious scientist will answer: "The side facing us is red."

This caution is manifested in many ways. For example, we hang onto our theories; we do not discard them readily. If a new observation suggests that a long-held, well-documented theory is wrong, we do not immediately discard that theory. We scrutinize the new information to assure ourselves that it is indeed properly interpreted—in other words, real. If the observation stands up to critical inspection, then we must accept it, in which case our theory is modified to incorporate the new data. But if that modification cannot happen, the theory must be discarded. This is why we describe our methods in great detail in our scientific reports. We need to provide our colleagues with the information necessary to reproduce our findings. If others cannot obtain the same results, we don't believe those results.

In November of 1977, a geological expedition using the submersible *Alvin* dived 8,000 feet into the blackness of the Galapagos rift in the eastern Pacific Ocean. The scientists were looking to see if hot springs, also known as hydrothermal vents, existed on the sea floor where the Earth's great plates were being pulled apart. They found their vents but were blown away to discover the vents were covered in foot-long clams, white crabs, and three-foot tube worms growing on top of one another in dense profusion. The vents were like rich oases in the otherwise sparsely populated sea floor, where the only available energy was the small amount of dead material that remained intact after slowly sinking from the sunlit ocean surface a mile-and-a-half above. There were no biologists on that expedition, but a year later, a team of biologists did visit the site. In time, it became clear that this newly discovered ecosystem was entirely dependent on chemosynthesis.

We are all familiar with photosynthesis, the process by which plants and some bacteria capture the energy of light to produce the chemical energy in carbohydrates. These carbohydrates can be transformed into proteins, fats, and all the other complex biological molecules that exist in living organisms. Another energy source for the synthesis of carbohydrates resides in certain simple chemical compounds such as hydrogen sulfide, the gas that gives rotten eggs their aroma. It also gives the mud in salt marshes and other oxygen-free environments its characteristic smell. Chemosynthesis is the term that expresses the chemical source of energy in the production of carbohydrates, as opposed to the light source in photosynthesis.

One afternoon in October 1977, just a month before the *Alvin* expedition's discovery, I was explaining chemosynthesis in some detail to my Marine Biology class, at which point a student raised her hand and asked: "Are there any ecosystems based on chemosynthesis rather than photosynthesis." My answer was an emphatic "no," and I went on to explain that chemosynthesis only occurs in the absence of oxygen, which only happens in restricted locations such as the muds mentioned above. Those muds are free of oxygen because it is consumed in the decomposition of living things that were produced as a result of photosynthesis.

The following month, I was enthralled to learn of the spectacularly rich ecosystem surrounding hydrothermal vents in the dark depths of the sea. And I was humbled at the thought that I had said no such ecosystem existed. I guess I should have said: "As far as we know at the moment," no such ecosystem existed. I told this to my students—a teachable moment! As a scientist, I had always avoided making bold statements because I understood the tentative nature of our knowledge (recall "The side facing us is red"). My lapse underlined the significance of that stance. So, when you hear a scientist testifying to a government body or being interviewed by the media, and she sounds less definitive than you would like, she is just being a good scientist.

In 1980, I read about a new hypothesis suggesting that the impact of a massive asteroid caused the extinction of dinosaurs. I didn't believe it. I had been influenced by the suggestion of the eighteenth-century Scottish geologist Sir Thomas Lyell, who proposed the concept of uniformitarianism, the idea that all the observed physical features of the Earth could be explained by processes occurring on the Earth today. A striking feature like the Grand Canyon was not caused by some single cataclysmic flood but by the observable slow year-in, year-out incremental erosion over a very long time. In fact, it was this idea that inspired Charles Darwin to conceive of natural selection as the creative force in evolution. Tiny incremental changes can generate large differences if they occur over a long enough time; a fish can become a human. Uniformitarianism was very well documented in both geology and biology. The invention of radioisotope dating of rocks truly solidified the idea by demonstrating the great age of the Earth, thus giving adequate time for small changes to accumulate and produce large transformations. So, my thinking went, slow climate change or some deadly disease must have caused the dinosaurs to disappear. The simple idea that an asteroid hit the Earth, causing cataclysmic destruction, was too easy.

I then learned about the evidence of an asteroid impact. There is a distinct layer in the rocks at the time the dinosaurs went extinct. This layer contains high concentrations of iridium, a rare element on Earth but more abundant on asteroids. It also contains shocked quartz, created by pulses of great pressure, and small glass spheres generated when rock is melted by impact and sprayed as tiny droplets into the air. The purported impact crater has been identified at Chicxulub on the Yucatan Peninsula of Mexico. The concentration of shocked quartz and glass spheres is greatest near the Yucatan and gradually becomes reduced with distance. The thickness of the distinct rock layer similarly decreases with distance.

How could I ignore such strong evidence? My long-held belief in uniformitarianism was not wrong. Uniformitarianism just is not universal. Unique catastrophic events *also* played a role in the Earth's history.

My conclusion on another big topic, global warming, also changed with time, but in a different way. Since the late 1960s, I have entertained the idea that human activity can cause global warming. We had data showing that the carbon dioxide levels in the atmosphere were increasing, and it was known that carbon dioxide is a strong absorber of infrared radiation, the form in which heat escapes the Earth. The resultant warming was a hypothesis then. It took thirty years before I accepted it as fact. In this case, we were not assessing a static situation. We were accumulating data about a phenomenon that was in a state of flux. But the real problem was that the year-to-year fluctuation of random weather patterns masked the slowly increasing average temperature. Brigid would say: "Global warming is already happening." I would counter: "We can't be certain of that yet," usually in the context of policy discussions and what the government should do to ameliorate the problem. My point was the country should act to reduce our carbon release even if global warming was not yet certain because there were so many other upsides to that. In time, the number of sequential hottest years on record finally gave me the statistically significant result that sealed the case.

The irony is that this conservative process results in the greatest changes in our perspective: the Earth revolves around the sun; humans evolved from other animal forms; matter and energy are different manifestations of the same thing. A more open process might proceed more rapidly, and that may be satisfying, but it would lead us down many false paths. The capacity of the scientific process to completely change the view of our place in the universe comes precisely from the conservative nature of science. Conclusions come after much accumulation of data and critical examination of all alternatives, but when we accept them, it is with great confidence in their reality. This idea is beautifully summed up in the unattributed quote: "Science is like magic, but real."

My Father's Legacy

My lawn on Cape Cod is a limited yellow-green area around the house consisting of mowed grass and assorted other plants. I won't call them weeds. Breaking the monotony is the well pit—a deep hole containing a pump, pressure tank, and filtration unit. That hole is topped by a gray weather-worn wood cover surrounded by a mini wall consisting of a couple of layers of red brick. The bricks form a U-shaped structure enclosing a bed of daylilies that grow between the grass and the pit. By 2020, frost heaves over the years had loosened and jumbled these bricks. Some lay on their sides with chunks of mortar attached to them. Others fell off onto the grass, and I loosely placed them back on top of the still-secure bricks. I had been meaning to repair this mess for several years, and with the extra time I had due to retirement and Covid-19 isolation, I finally got to it.

Working with bricks has special meaning to me because my father was a bricklayer. His daily handling of bricks and cement blocks gave him thick, strong fingers and muscular forearms. At the dinner table, a glass would sometimes slip from his hand because his fingerprints were worn smooth by the rough masonry. He was fair-skinned, and instead of tanning, he just developed a permanent redness as he worked exposed to the full sun in the summertime. We didn't use sunblock back then.

I did not expect it, but repairing the bricks around the well pit transported me into that past. I recalled how my father handled bricks and mortar with such aplomb; he could have done it in his sleep. These memories were stirred by the tools I used, his tools, that I had inherited: trowels, a hammer, jointers, and a long wooden level, all worn by years of use. As I heard the scraping of tools against the mortar and smelled its slightly caustic odor, my father was there—in me, beside me, around me. It wasn't his spirit or simply his essence. He was physically present in his worn khaki work pants and clean white tee shirt, big red arms working the mortar into the right texture and slathering it onto bricks. And I was a little boy watching him work, not being impressed; it was just the way things were. It's simply what my dad did.

So, I mixed mortar and laid some bricks, not with the smooth, practiced movements of my dad, but in an awkward, hesitant manner that got the job done very inefficiently. Mortar is dense, and its weight quickly tired the muscles in my skinny forearm. I was essentially playing. Unlike him, I wasn't driven to lay the most bricks in the least time, almost like a machine.

Bricks provide strong, weather-resistant structure, but they are nothing without that amazing substance: mortar. Mortar is simply cement and sand mixed with water, but it has a schizophrenic nature. While still wet, it can be solid and supportive, or if shaken, it can be fluid and flow. An example of this is the liquefaction of wet soils during earthquakes. In certain locations, buildings and elevated highways that have been stable for many decades suddenly collapse when the shaking turns the wet earth into a non-supportive soup. This is what happened in the San Francisco earthquake of 1989 when elevated roadways collapsed, causing numerous injuries and fatalities.

The rosy razorfish, a kind of wrasse that lives on sand bottoms near Caribbean coral reefs, provides my favorite example of the phenomenon, however. It is an elongated, six-inch fish with a slim body and a narrow ridge along its snout. While scuba diving, I saw these guys watching me as they hovered a few inches above the bottom. If I got too close, they plunged headfirst into the sand and disappeared. By wriggling their bodies, they liquefied the sand and were able to swim into it, only to be embedded in firm sand when they stopped moving. I never waited to see them emerge, but doubtless, they exited by quivering their bodies and swimming out of the sand.

This property of mortar is critical to laying bricks. I used the trowel to mix the mortar, loosening it so I could easily slip a portion onto the face of the trowel. I slopped the mortar onto the existing bed of bricks, lifted another glob onto the trowel, and slid it onto the end of a brick I was holding. Then, I placed the brick onto the mortar bed with the mortared end against the end of an already seated brick. Here's where the schizophrenic nature of mortar comes into play. By tapping on the brick with the handle end of the trowel, I liquefied the mortar. The brick sunk into position, and the mortar infiltrated the rough surface of the brick, forming a tight bond. It then solidified, and the brick remained securely in place. Finally, by rubbing the joints with the cylindrical jointer tool, I liquefied the surface of the mortar again and could shape it into a smooth, clean joint. Without this property, it would be very difficult to position bricks with precision, and they would not be tightly bonded.

I always knew about this simple but special dual nature of mortar in some vague way, but it wasn't until I laid these bricks in my backyard that I came to understand it fully. And my dad was there, guiding me along.

Repairing the well wall was not the first time I used the construction skills inherited from my father, and I don't mean only in building blenny homes. Many years before, while he was still alive, I designed and built a solar greenhouse. This was in the early 1980s during the first wave of national energy awareness following the 1973 and 1979 energy crises. Karen and I purchased our house in Ossining and noted that the south-facing wall was the only one without windows. My gears began turning.

I was soon off to spend a year on St. Croix, and I carried several books on solar design with me. During my free periods, I read them voraciously—I really enjoyed the technical aspects. I have mused several times that had my early exposures been different, I could have become an architect. In any case, I began drawing detailed plans for a half-underground lean-to greenhouse to sit against the south wall of our house. I really didn't think I would actually build it and referred to the project as my "pipe dream."

One of Jimmy Carter's responses to the 1979 energy crises was to implement significant tax credits for the use of energy-saving technologies, including solar. My greenhouse dreams became feasible. I began to build. One aspect of solar design is the incorporation of thermal mass. By having a dense material that absorbs the excess heat during the day and releases it at night, I could avoid extreme temperature shifts. My plan was to have eight tons of fist-size stones under the floor through which a fan could blow air when the greenhouse got hot. I hired a man with a backhoe to dig a foundation six feet deep. During the digging, he exposed a large boulder and left it embedded in the soil. I then built frames and poured a concrete footing, incorporating the boulder that projected above the flat surface. I laid a cement block foundation, being careful to keep the layers level as I built around the boulder. On visiting us in the fall, my father praised me on how I had incorporated the boulder while keeping the block courses perfectly level, a welcome compliment from a professional—not any professional, my dad. He rarely gave compliments.

After that first summer of work, the foundation was complete, and the greenhouse consisted of a cement block-lined pit projecting from the foundation of our house. It was fourteen feet long, eight feet wide, and six feet deep. One Saturday morning in October 1978, I saw a scraggly gray furry mass on the bare earth floor. I placed a ladder in the hole and stepped down to investigate. It was a raccoon that had fallen in and died, probably in a horrible state of starvation or dehydration. A feeling of guilt came over me as I removed the body. I then leaned a pole against one wall to provide a way out for any other unfortunate animal. The following month, the family raked leaves together in our annual shared chore. We completely filled the pit with leaves to prevent the ground below from freezing. Frost heaves can crack a foundation, and I was concerned not only for the new structure I had built but also for the foundation of our house that I had exposed.

The next summer, I took a sledgehammer and rammed a hole through the house foundation, carefully framed that opening, and installed a glass door from the basement to the greenhouse. After I built stone steps three feet up to the anticipated floor level, a large dump truck backed to the edge of the pit, tilted its bed, and with much rumbling and clacking, eight tons of fist-sized rocks slid in and filled the pit to the level of the top step. I built a wood frame on the block foundation, installed siding on the east and west walls, and put shingles on the roof. Everything was heavily insulated. On the south side, I installed large thermopane glass windows angled to be perpendicular to the noon sun on December 21, thus maximizing the solar gain in winter.

To complete the transformation, I had a large picture window installed on the south side of the house as well.

The half-underground solar greenhouse. Photo by author.

How did the greenhouse perform? On a 3-degree Fahrenheit day, when clouds prevented any heat gain from the sun, the low temperature in the greenhouse was 48 degrees Fahrenheit. The secret was the effective insulation, glass only on the south-facing wall, the north wall against the house, the bottom half underground, and the stone thermal mass. On sunny days, the blower pushed air through the stones under the floor, and once the stones were warm, the excess heat was pumped into the house. In time, the greenhouse was filled with a great variety of plants, and I enjoyed walking in and seeing what was new as they grew and flowered. One morning, I was assaulted by the smell of rotting flesh, not unlike the odor of the poor raccoon that had succumbed there a couple of years earlier. A ton of flies were buzzing about one corner by a starfish flower (*Stapelia*), a cactus-like plant from southern Africa. Its large flower had opened overnight and emitted that rotten smell. Nature has no esthetic sense, and while bees and butterflies often pollinate sweet-smelling flowers, carrion-loving flies will do the trick, too.

I felt a sense of great satisfaction that I had designed and built this structure. This physical work provided a different sense of gratification from my intellectual work. Seeing many hours at the computer result in a publication is very rewarding, but it is an abstract accomplishment. Seeing the greenhouse standing there and performing as expected is a concrete accomplishment. When doing any chore that involved building something—a bookshelf, an end table, or a fence around the vegetable garden—I stop and gaze at my work from time to time for several days after completion and think: *That looks great, and I did it.* I don't need anyone else to tell me that it is good work. The reward is immediate. Putting a research paper out there is different. The satisfaction comes from how it is received and how often it is referenced in other people's work. The gratification comes slowly and in a delayed fashion, often spread over many years.

His Demise

In August 1990, on a visit home, I walked with my dad beside the house, and he brushed his upper left arm against the rough cement foundation. He bled profusely from a dark, irregular mole. I said: "You'd better get that looked at by a doctor." He didn't take care of it until late October, when a biopsy was taken. At the time, he said: "If it's cancer, I don't want to know." It was melanoma. He had surgery, but the cancer had already metastasized, and he went home to spend his last days.

I spent as much time with him as I could that spring. I watched a strong, vigorous man deteriorate. He had no interest in looking at the striking tulip flowers he had planted in the fall. Mum did what she could, supporting him as he walked to the bathroom. He fell a couple of times, and she had to call a friend to help get him up. His arms and legs got heavier with retained fluid, and as he lost strength, he couldn't lift them anymore. He was bedridden and had hallucinations, seeing little dwarf men running around the room. Giving him a sponge bath in bed, a visiting nurse said: "I'll leave your private parts for your wife." His reply was: "I don't have private parts anymore."

In June 1991, after spending two weeks in the house being with him and helping Mum, I had to return to Ossining to complete my year-end student evaluations. Before I left, I said to him: "I love you." It was the first time I'd ever said it, and he answered: "I love you all." Those were the last words he said to me and the only time I ever heard him use the word "love," and even then, he could not directly say "I love you." This emotional expression, even if limited, was so striking, even more than the awful physical transformations of his body.

Two hours after arriving home, I got a phone call from the nurse saying that he was near the end, unconscious and presenting the death rattle. I jumped in the car and drove another seven hours, arriving in the early hours before dawn. I sat by him, with Mum and Wendy in the room, as he lay there unconscious with that horrible gurgling breathing of fluid filling his throat. Less than an hour after my return, he took his last breath, a deep inhale . . . and nothing.

I was in shock. We had tea at the kitchen table as the sun rose. Nobody spoke. Wendy had the presence of mind to make the call and have his body removed. I sat on the balcony and watched the morticians wheel his covered body down the front path. A woman neighbor, sitting on her balcony, lurched forward to look as she saw what was happening. The morticians slid the bundle into a slot in their truck and latched the door. Off they drove. I was tired, drained, flattened. Time stopped. The world stopped.

Many times, I have passed a slowly moving funeral procession on the road—a black hearse followed by a string of cars, each with its headlights on to indicate it was part of the procession. This time, on a bright, sunny day in June 1991, I was part of a line of cars on Sherbrooke Street escorting my father to his resting place. Mum and he had purchased a plot in a cemetery on Mount Royal, the large hill in the center of Montreal. When they'd purchased it, Dad said: "One day, I'll be looking down on you schleppers." On this day, I stood beside a hole in the ground with my father's coffin at the bottom. I picked up a handful of soil and swung my hand to release it, the first step in his interment. But I couldn't let go. It felt too final. I tried again, and the soil dropped onto the polished wood coffin with a crumbly, grating sound. The pristine surface was now marred with dirt. Six feet under.

Up in Smoke

My dad's occupation as a bricklayer in Montreal in the 1950s presented a problem. You cannot lay bricks in winter weather because the mortar freezes before it cures. Back then, we didn't have today's portable heaters and enclosures, so construction stopped in the winter, and the workers were laid off without compensation. I always knew when the tight winter period was upon us because my parents would start rolling their own cigarettes. Mum had a little metal machine the size of a wallet. She placed the paper and loose tobacco inside, pulled on some mechanism, and out popped a perfectly formed cigarette. Dad was cool and could roll a cigarette in one hand, cowboy style.

On fishing trips, we walked along trails in the forest to reach the lake. A frequent sound was *hack-hack, hack-hack,* followed by my dad's friend Bill saying: "Your father's coming." Bill told me he always knew where my father was because he could hear his smoker's cough through the trees. Bill smoked, too. He had a habit of puffing out his cheeks as he took a drag and then filling his lungs. As I grew up, just about every adult around me was a smoker.

I have always hated smoking. The smoke was choking, and the ashtrays stunk. I have a strong memory of the smell in the tiny bathroom as my mother prepared to go out. She would be looking in the mirror, applying Noxzema cream, with smoke rising from the ashtray beside her, creating a nose-battering combination of menthol, camphor, and eucalyptus finished with a heavy dose of cigarette smoke. When I complained, Mum said: "You'll probably end up marrying a girl who smokes like a chimney." Little did she know what a turnoff it was to me. It's a wonder I haven't developed lung cancer after inhaling so much second-hand smoke in my life: in the car with two smoking parents and the windows shut tight against the biting Montreal winter; during parties in the basement playroom with my parents, aunt, and uncle all smoking and barely being able to see across the room through the haze; and in various meetings as a professor, with everyone but me puffing away. When I was on The Committee for Student

Work, the Dean, the Associate Dean, three other faculty members, and two student representatives all lit up. As they sat around the table discussing cases, I participated at a distance, seated on the sill with the window cracked open to the crisp, fresh winter air.

Times have changed now, and it's the smokers that you see huddled against the cold in building doorways, satisfying their addiction while paradoxically getting fresh air as well. Humans have a remarkable capacity for ignoring unwanted information. We all do it. But the extended time it took for smoking to become unacceptable in the face of the evidence is remarkable. Of course, the tobacco companies sowed disinformation just as the fossil fuel companies do today, and this worked especially well when falling on willing ears. However, my problem with smoking was not its health effects; I just did not like the smell and the choking feeling of inhaling smoke. I believe this is an adaptive response we all have. Novice smokers have to learn to tolerate the irritants they inhale. Even as an adult, the thing that kept me from enjoying a communal joint was the dislike of inhaling smoke.

My father had a triple bypass at the age of 67. He was frightened when he went into surgery, and he told my sister Wendy: "If I don't make it, I want you to have the car." He was overweight and did not exercise, but smoking was probably a contributing factor as well. Warfarin, or as we called it, "rat poison," was among the pills he took daily after surgery. It was prescribed to slow the clotting time of his blood and prevent a heart attack or stroke. Because nicotine enhances the clotting of blood, his doctors told him to quit smoking. We thought he had, but I later discovered that he was sneaking cigarettes in the basement. As far as I know, his melanoma was unrelated to smoking.

The poisonous effects of tobacco smoke were demonstrated to my parents and me in a striking way when I was just seven or eight. I would catch blue bottle flies in jars, and sometimes, my mother would blow a puff of smoke from her cigarette into the jar. The fly would buzz about the jar only to fall to the bottom, twitching and twitching as it lay there upside down until its six legs folded over its body and all was still. Sometimes, we would watch two or three flies die together. I cannot fathom how my mother could keep smoking after seeing what the contents of her lungs did to a fly. When I was older and told her how dangerous smoking was, she always protested that she did not inhale. Of course she did. The disconnect between her behavior and health was underlined by her casually referring to cigarettes as "cancer sticks."

My mum, a vigorous, energetic woman, died of pancreatic cancer in 2004. She had been living with Wendy in Markham, near Toronto, while I was 450 miles away in Ossining. On my visits, I watched her wither away, her legs oozing a foul-smelling fluid, yet she remained in good spirits. At least she pretended to be in good spirits because her main concern was how my sister and I were handling her demise.

Before Mum's illness greatly incapacitated her, we had a family get-together in Sedona, Arizona—Mum, Wendy, Serena, Justin, and me. I rented a nice house, and we set Mum up in the master bedroom that had a large picture window facing the beautiful red rock mountain peaks. We went for drives and walks and window-shopped downtown, and I fully appreciated the nurturing selflessness and warmth of Mum. As a child, I had taken cues from my chauvinistic dad and treated her as a lesser person, not as smart or sophisticated as us (which we weren't). No more!

The Sedona City website states that: "Sedona vortexes (the proper grammatical form 'vortices' is rarely used) are thought to be swirling centers of energy that are conducive to healing, meditation, and self-exploration. These are places where the Earth seems especially alive with energy. Many people feel inspired, recharged, or uplifted after visiting a vortex."[49]

After her return to Markham, Mum had a doctor's visit and was told that the tumor had disappeared! Did the "swirling centers of energy" heal her? She had a year of relatively good health before the cancer returned with a vengeance. Some New Age believers would say that she would have remained healthy if she had stayed in Sedona. It is possible the trip and our family warmth put her in a positive mental state, which may have resulted in her immune system functioning extra well. That could have reduced the tumor. But the healing powers of some kind of swirling "energy?" Not for a minute did I entertain that idea.

Early in her illness, when I searched the internet to understand the disease, I discovered that the only environmental factor associated with pancreatic cancer was smoking. I did not tell my mother that.

From the Ashes

On a clear fall day in 2004, the sun shining weakly, I finished my gardening chores at Brigid's house and went inside. I was ready to put my feet up for a little while, but as soon as I opened the door, I noticed the sharp smell of overheated electronics. I immediately went into investigation mode, checking all TVs, computers, printers, and modems in the house, but none of them were emitting an odor. Opening the door to the basement, I was hit by a wave of intense, acrid smoke. I didn't go down but ran around to the outside entrance to the basement and tentatively eased my way in. The smoke was choking me, so I held my breath and went further for a quick look. I couldn't see more than six feet through my burning eyes. As I approached my workbench, I could make out its companion, a red plastic trash can. Flames were shooting out of it—not little gentle flames but vigorous, lively flames reaching three feet high. That's why I thought I'd smelled overheated electronics; it was really the smell of burning plastic, with its potentially toxic fumes. I ran back outside to breathe fresh, sweet air before heading to the hose rack on the side of the house, where I turned the water on and uncoiled the hose. Taking another deep breath, I went back down, dragging the hose with me. I hosed the fire for as long as my breath lasted and ran back outside for air. After repeating this sequence two more times, the fire was reduced to a gentle thread of steam rising into the pungent, smoky atmosphere.

Now what? Oh my God. This fire was my fault. I knew better, but there it was—the result of my carelessness. Before the gardening, I had refinished the beautiful mahogany surface of the front porch with Australian timber oil. I had rubbed the oil into the wood with rags and, when finished, disposed of them in the plastic trash can. Ever since I was a kid, I had heard that you should not dispose of paint rags in a garbage can, yet that's exactly what I had done. The oil in the rags had oxidized, giving off heat in the process. The heat had built up until the ignition temperature of the saturated rags was reached, and they'd burst into flame—spontaneous combustion. It didn't help that I had previously disposed of wood scraps in the can. I couldn't have created a better demonstration of the danger of "oily rags in a garbage can" if I had purposely tried.

I had a long list of chores to do that Saturday, so once the excitement was over, I moved on to cleaning out the gutters. I was standing on the roof over the kitchen when Brigid came home from a meeting. As she walked around the corner of the house, she called up: "Thank you for taking care of my house." I called down: "I almost burned it down." That piqued her curiosity, and I explained in some detail what had happened. The house was a three-story Victorian built in 1893—all wood construction with wood siding—old, dry wood. A fire in the basement is the most efficient way to set a whole house ablaze because the rising flames would ignite the upper floors in no time. Had I not been home, the house would have become a conflagration and burned down to the stone foundation. I really did come close to totally destroying that house. Unlike Miss Patton, who lost her old wooden house to fire, I was lucky to be there to put out the flames.

Later that afternoon, I cleaned up the mess. The basement still had that unpleasant smoke smell, so I opened all the windows to air it out. One leg of the workbench was charred but still strong enough to do its job. I was left with the remains of the trash can. It had been reduced to a hardened puddle of red plastic. The combustibles, including those soaked rags, had been totally consumed, leaving a tangle of wires embedded in the plastic.

The next day, the assemblage looked like a piece of art to me, so I decided to make the most of it. I hosed it down to remove the ashes and other loose pieces. After letting the remains dry out overnight, I sprayed them with a clear acrylic sealer. The plastic now appeared shiny, like something brand new. It was an ironic, bright fire engine red. The finishing touch to the work was a wire brush that must have fallen off the workbench during the conflagration. The brush's head was embedded in the plastic, and its handle, with a hole for hanging, stuck out of the side. I now had a piece of art with its own built-in hanger. We can call the style "inadvertent art." It now hangs on the wall above my new workbench in Truro as a stern reminder to remain humble and to work with care and thoughtfulness.

Melted red trash can – Inadvertent Art.
Photo by author.

Shifting Baselines

In 2001, I read a paper in *Science* magazine entitled "Historical Overfishing and the Recent Collapse of Coastal Ecosystems."[50] That article introduced me to the concept of shifting baselines, the idea that without strong historical information, we tend to measure environmental change in relation to our personal experiences. In other words, our first encounter with a habitat is taken as the basis for measuring future change in that habitat. We are generally unaware that we are seeing an already degraded habitat, not one in a pristine state. How would we know? By 2001, the coral reefs of the western Atlantic had been severely degraded from what I saw at Looe Key in 1969, and that reef had been far from pristine. Today, tourists visit coral reefs and marvel at their beauty, having little idea of what I once knew. I don't wish to minimize their wonderful experiences. Even though the corals are not what they were, the fish are still an eyeful, in spite of their reduced populations.

I visited the Discovery Bay Marine Lab in Jamaica in 1978. My host was Les Kaufman, who at the time was known for his discovery that three-spot damselfish farm their algal gardens by killing live coral to create a footing for the algae.[51] They subsequently weed out the unpalatable species by biting off pieces of distasteful algae, swimming to the edges of their territories, and spitting them out. Although the Jamaican reefs were famously overfished, the corals were spectacular. The shallow forereef was covered in great stands of the massive branching elkhorn coral, a tree-like species that can reach eight feet in height. It is arguably the most spectacular coral in the world, with spreading branches the color of brown mustard, a dazzling sight against the deep blue of the clear Caribbean water, especially on sunny days when the bouncing prisms of waves bathed the corals in flashing specks of light as if lit by a ballroom mirror rotating at double speed. The growth was so prolific that the surface area of the overlapping branches far exceeded the area of the reef surface they covered. This abundance was not unusual in the Caribbean at the time. A snorkeler could see endless forests of these marine trees, growing and growing, laying down their limestone skeletons, building the foundations of the fringing reefs that protect the islands of the Caribbean from the perpetual grinding of the Atlantic swells and the periodic hammering of hurricanes.

The elkhorn coral on the reef at Teague Bay in St. Croix was similarly exuberant when I first visited in 1977. By 1992, the area looked like a parking lot—a flat reef surface covered in white rubble, white being the color of the carbonate skeletons of dead coral.[52] Les Kaufman's three-spot damselfish no longer needed to kill coral to provide a surface for algae to grow. Indeed, the coral reef had become an algal reef. In 2008, the International Union for the Conservation of Nature listed elkhorn coral on its "critically endangered" list, the highest category before "extinct."

The causes of coral decline worldwide are the subject of much discussion. Clearly, elevated temperature due to global climate change is a major cause, but there are many other contributing factors, such as disease, acidity, excessive nutrients, overfishing, and sedimentation. In the reef science community, there is little question about the negative effects of these elements. Scientists' remaining questions relate to those elements' relative contributions to coral decline, contributions that vary from reef to reef. It is unlikely that any widespread corals will actually go extinct. What will probably happen is that coral reefs will disappear. These beautiful, structurally complex structures will be replaced by simplified pavements harboring small scattered corals. Undoubtedly, many associated organisms will disappear, many that we will never know even existed. If I am the only person to have seen one coral clingfish, how many other species have been seen by nobody?

Because corals require very clean water and they are living close to their maximum temperature tolerance, coral reefs have been compared to "canaries in the coal mine" with respect to global climate change. The only ecosystems rivaling coral reefs for vulnerability occur at the opposite end of Earth's temperature range: the Arctic Ocean and nearby permafrost tundra. Coral reefs are the only biological structures visible from outer space without a telephoto lens. The area of Australia's Great Barrier Reef is 134,634 square miles. A healthy coral reef is a robust structure, resisting the pounding of ocean swells and recovering quickly from hurricanes and cyclones. In spite of this, they are the most sensitive systems to global warming. What an irony that the largest, most majestic biological structures on Earth are the most vulnerable.

My recollections of reefs from my first dives are similar to the ones of other divers my age—we see today's reefs and pine for what is gone. While nobody recalls the pirate days when the Caribbean was awash in sea turtles, resulting in so many locations being named "Tortuga," many divers recall the lush coral gardens from the 60s and 70s. The baselines are shifting at a greater and greater rate, and this acceleration is truly frightening.

The "Good" Old Days

My father and his friends, Bill, Joe, and Dez, took a fishing trip every Labor Day weekend, and when I was old enough, I went along. We always went to the same place, an old logging camp in Parc La Verendrye in central Quebec. The logging had occurred because the land was to be flooded for a gigantic hydroelectric project. The flooding created a very large lake with serpentine shores marked by bright white paper birch trees backed by deep green spruce and fir, all reflected in the dark water that bore tremendous populations of walleye and northern pike. Fishing there came close to the proverbial shooting fish in a barrel.

We fished from boats that we had trailered up the 260 miles from our home in Montreal. After a morning of fishing, we would pull our two boats ashore, make a fire, and have lunch. One notable day, the air was hot and still; the sun was beating down. The atmosphere was oppressive, and my father and his friends took their usual post-lunch nap. They were spread out among the bleached snags of tree roots that had washed out of the lake bottom and littered the shore. Their hats were over their eyes. Their breathing was heavy. Joe had his fork still in his hand resting on his big belly. Being fourteen at the time, had too much energy to stop moving like the old guys, so I decided to go for a cooling swim.

My dad (foreground) and Joe after lunch at Parc La Vérendrye, Quebec. Photo by author.

I swam along the surface and then, as I always do, dived towards the bottom. When I opened my eyes, it was pitch black down there. Well, it felt pitch black but was really a very, very dark chestnut brown. The key point is that it scared the shit out of me. I shot up to the surface, never to submerge again that day. I had known that the water was brownish but had not anticipated that it would absorb so much light in only about six feet. A high concentration of tannins from so many submerged trees had dissolved in the water. After I emerged and the old guys awoke, I dipped a glass into the lake for a drink. The water was teeming with ostracods, spherical swimming creatures slightly larger than a printed period. I emptied the glass and dipped it again in another spot with the same result, so I drank it, ostracods and all—my first protein drink, and fresh protein at that.

The trips happened well before we were aware of the danger of the protozoan parasite *Giardia,* which is so common in all freshwaters. It causes a month of diarrhea, stomach cramps, dehydration, and fatigue. There were other unknown dangers at the time. With the damming of the river and the thousands of acres of flooded forest, a tremendous amount of nutrients ended up in the water. This supported algal growth, which in turn supported large zooplankton populations, including the ostracods I drank. These were consumed by small fish, which in turn were consumed by larger fish, and so on through a couple more steps until reaching the large walleye and northern pike that drew us there. A feature of such food chains is that some chemicals are concentrated, reaching higher levels with each step up the chain. It turns out that the flooding released mercury from the soil, and in its organic form, methylmercury concentrates in the fauna. We have since learned that those northern pike had mercury levels elevated seven times above normal in nearby reservoirs and that indigenous Cree people, who depend to a large degree on fishing, had the highest measured methylmercury concentration of all Canadian First Nations, fifteen percent of whom exceeded the established safe norm. So, going "up north" to the wild, seemingly clean lakes and forest was not the healthy activity we'd thought it was. The so-called "good old days" represent not only our selective memories but also our state of ignorance back then.

A few years later, I saw this ignorance compounded by heartlessness. Hydro-Québec, the state-owned electric company, had plans to flood another large area in central Quebec in what was called the James

Bay Project. The hydroelectric scheme covered an area the size of the state of New York and was populated by Cree and Inuit, living largely by hunting and fishing. These residents fought the government effectively and received a large financial settlement, which was hailed in the press as a victory for the native peoples. My thought at the time was: *yes, a victory, as seen by those who measure value in dollars*. But the people did not get what they most wanted—to continue living in the manner of their ancestors on the land they had occupied for countless generations. This option was never up for negotiation.

Loss and More Loss

I can't remember the last time I saw a mayfly flitting about on its long diaphanous wings. At certain times in the early summer when I was a kid, mayflies and shadflies would emerge from the ponds and lakes in tremendous clouds. The air was alive with insects. As my father drove his shiny, new, green 1954 Chevy Belair along any country road, the windshield became smattered with squashed bugs of all sorts—a large assortment of moths at night and an assemblage of flies, beetles, and butterflies during the day. Occasionally there was a thud as he ran into an extra-large insect, maybe a carpenter bee or a junebug or a big grasshopper. If he turned on the windshield wipers, the splats would spread out into greasy streaks. When we drove into the sun during the day or toward the headlights of an oncoming car at night, it was almost impossible to see. Every gas station had buckets of soapy water for patrons to clean the bugs off their windshields. This was more than a courtesy. It was a safety issue.

In a carefully controlled study in Denmark measuring the number of dead insects on car windscreens on two stretches of road between 1997 and 2017, Anders Møller of the Centre National de la Recherche Scientifique in Paris found 80% and 97% declines in insects.[53] This result, along with many other sources of data, has prompted some environmentalists to declare that we are experiencing an insect apocalypse. The problem is even worse than any of us think because we are measuring the decline from an already decimated starting point. There's no way that the 1997 baseline in the Denmark study represents the pristine population density of insects, so eighty percent is a gross underestimation of insect decline. It's a case of shifting baselines. Virtually all our studies of environmental decline start with an already degraded state. I do, however, imagine that the insect populations in 1954, when my eight-year-old self rode in my father's car, might have been close to undiminished. DDT was a recent discovery and was not yet widely used, and many other horrible pesticides were not yet invented. Rachel Carson's book, compellingly titled *Silent Spring*,[54] envisioned a world with few birds because of DDT poisoning. And where did birds pick up DDT? From insects and other organisms in the food chain. Ospreys and brown pelicans flirted with extinction by eating DDT-laced fish.

Ecologists frequently lament the loss of species because of reduced "ecosystem services," the idea that many species provide value, such as pollination by bees or mosquito control by bats. Because we have such a poor understanding of the role of most species in the functioning of ecosystems, ecologists caution about allowing any species to go extinct. This idea was persuasively expressed by Aldo Leopold in his ground-breaking book, *A Sand County Almanac*: "To keep every cog and wheel is the first precaution of intelligent tinkering."[55] Many people point to potential economic value as the reason to prevent extinction as much as possible. An often-quoted example of this is the rosy periwinkle, a plant endangered in the wild by deforestation in Madagascar. It is the source of a potent anti-cancer drug that can reduce the mortality from childhood leukemia from 90% to 10%.

But the rosy periwinkle is easily cultivated. So why should we worry about its extinction in the wild? If economic value is the main reason for saving species, then what do we do when we can achieve that value differently? For example, because of the reduced populations of wild introduced honeybees and our own native bees, a new industry has popped up. Honeybee hives are now transported between states en masse on flatbed semi-trailer trucks to almond orchards, blueberry fields, and many other crops that need pollination by insects. Even more unsettling, Eijiro Miyako, a researcher at Japan's National Institute of Advanced Industrial Science and Technology, has developed bee-size drones that can pollinate artificially.[56] So, do we need bees? The pragmatic economic argument for saving species is compelling to legislators, most of whom have no sense of ecosystems as highly interactive assemblages of co-dependent species. If a species' known economic value can be achieved by other means, they may see no reason to work to save it.

Winding Down

After forty years of teaching at Sarah Lawrence College, I still enjoyed being in class and sharing my knowledge with students. Even more than the information, I purposely conveyed my fascination with the natural world and the excitement of evaluating new hypotheses and how they furthered our understanding. No, it wasn't the class time that prompted my retirement; it was the homework. I just didn't have the energy to do all the reading and class preparation every night. And without being thoroughly prepared, I just didn't feel right. Some people can "wing it." I can't. I have done it successfully a couple of times when emergencies prevented my preparation, but I am otherwise unable to let myself do that. A guest speaker once described lecturing as the process by which information gets from the professor's notes into the student's notebooks without passing through either's mind. That, in fact, was much of my personal undergraduate experience. I could never replicate that. It was time to go.

As my retirement approached, I spent a great amount of time disassembling my office, the third one I'd occupied during my teaching career. I had been in this one for twenty years. While bookshelves and desk drawers were filled largely with materials I had been using recently, there were also a few items remaining from my student days, such as the classic book *Life of Vertebrates*[57] by JZ Young and the influential book *Genetics and the Origin of Species*[58] by Theodosius Dobzhansky, the latter underlined almost completely. I also had a slide rule from 1965 and a l x 5 mm piece of chromium sealed in a piece of glass, a memento of my doctoral research when I'd had used tiny chromium rods bombarded with positrons and converted to radioactive manganese-52 to locate the toads I was studying.

But the most nostalgia and historical insight emerged from the three four-drawer gray metal filing cabinets. They contained the records of all the students I advised over the years. They contained the notes of all the courses I taught. They contained all the memos I wrote and many of the ones I received. They contained the minutes of the many committee meetings I attended. And they contained many of the papers, class notes, and lab reports that I produced as a student. Two major themes emerged as I reviewed everything.

One, the detail and precision of my work fifty years previously as a student was much greater than that produced by the students I taught. My old lab notes were comprehensive, and the drawings were detailed and carefully executed. I loved that stuff and participated with energy and enthusiasm. I hate to say it, but much of the work my recent students produced was not up to that standard. I hope that's not a reflection of my teaching, but I suspect, to some extent, that it is. I don't take this entirely personally. I think that expectations have relaxed across the board. I have said many times to students that it doesn't matter if they don't draw well. The point of drawing is to make you see. And that's true, but if drawing is seen as a way of recording, then quality does count. However, if you are not intending to become a scientist and envision biology as one component of a liberal arts education, then it's the seeing, not the recording, that's important. Although their lab notes did not match mine, my students' willingness to participate in class discussions and the effectiveness of that participation greatly exceeded my experience as an undergraduate. Many of my students became doctors, and a few became professors, so something good was happening.

The second theme that emerged was that I was much more productive in my early years than in my later years. That's not to say I was a better teacher early on. I wasn't. But the number of memos I wrote and the number of committees I served on up until the last ten years was considerably greater. The energy I had in my early years is inconceivable to me now. My doctoral thesis was not quite completed when I began teaching. I finished it during the winter break of my first year and defended it during spring break. I had not taught before, so everything was new. I had no old notes to modify and update. All the labs were new. I worked nonstop, only taking Sunday afternoons off. I don't think this is especially unusual. In many fields and early in their careers, most people spend equal amounts of time working. I have strong nostalgic feelings looking at that from where I am now with less ability to focus deeply for long, needing occasional afternoon naps, and simply being less intensely motivated. From here, I look at that productivity and think: *Who was that guy?*

Recently, I have been going through boxes and boxes of 35 mm slides that I have taken over the years. Aside from all the shots of flora, fauna, landscapes, and artsy close-ups of bark, rotting wood, flowers, and insects, I have family and travel shots. There I was in the early years, skinny, dark brown, almost black hair, often without a shirt, engaged in adventurous activities. This young man was living life to the fullest and in the moment. It's me, but it's not me now, and thus, in some ways it is another person. I remember some of the times, but not all of them. I have changed so much, become more mature and less vigorous, but I still have the same love of Nature that I had then. I still love the same music: blues and 60s folk. The songs that most deeply affected me then still affect me deeply. John Lennon's "Imagine" still makes me pine for a perfect world. Some things don't change.

I was a good teacher. I loved my subject and taught with enthusiasm, but I depended on my students to keep me excited. Little did they know what an impact they had on me. If they nodded and asked probing questions, I was confident and put on a good performance, but if they were flat and unengaged, I would lose confidence, get nervous, and perform poorly, speaking in a stilted manner and forgetting some points. At least, this is what it felt like to me, but that may not have been the students' perception. More than once, after conducting what felt to me like an uninspired class, a student came up afterward and said: "That was the best class of the semester." Oh well.

In any case, I tried to be informal in class and to teach with levity. I introduced my personal experiences when appropriate and often projected 35 mm slides I had taken. I brought in bird bones, coral skeletons, sponges, mollusk shells, and an assortment of other things from my cabinet of curiosities. I stressed that biology was the study of living things, not words on a page. To that end, I would sometimes take a field trip at the beginning of a General Biology course, sit the students on the forest floor, and ask them to remain quiet for ten minutes. I then had them tell me what they'd observed. It may have been a bird that flew by or the gurgling of a nearby stream. Then, I would tell them what I had perceived, including a discussion of unseen processes going on in the soil, tree trunks, and the canopy above, and how they were all connected by conductive tissue running from the tree roots up to the leaves. I hoped to open their eyes and stimulate their curiosity about the living world around them.

I guess you could say I had a passion for biology that was underlined one May morning after graduation at Sarah Lawrence College. The commencement speaker was the novelist, Ann Patchett. Ann had taken my General Biology class twenty years earlier when she was an undergraduate. Michele Meyers, president of the college, and a few others were standing in her office making small talk when Ann volunteered that I was so fervent about biology that she recalled my tearing up when I'd once been talking about some fine point of cell structure. Everyone looked at me. I didn't know how to react and simply shrugged.

Over the years, I really did not understand the extent to which teaching was in my bones. I had always wanted to be a researcher and saw teaching, at least initially, as part of the package that comes with being a professor. However, during the first autumn after my retirement, I found myself on a rocky beach in California, and I picked up a piece of soft rock that was riddled with holes excavated by boring clams. My immediate thought was that I should bring this back to show to my students, but I suddenly realized I had no students. That was the first time I understood how much my identity was tied up with teaching more than anything. A hollowness descended upon me. I missed those fresh faces and the feeling of hearing their comments and sharing my knowledge. I missed my colleagues, too. The math and science faculty at SLC was a tight group that really worked well together. I knew that, but I had not understood how much they meant to me and how much being a respected member of a close social group gave me a sense of self-worth. Much as I thought of myself as an independent-minded scientist who happened to work at Sarah Lawrence College, I did not appreciate until then how much I had become enmeshed in the fabric of the school.

Once and Forever a Biologist

After many years of research on coral reef fishes, I decided to look for a project a little closer to home, something I could do on a regular basis without the burdensome logistic preparation for a field trip to the West Indies. This was to be my transition project, a research study to start before my retirement and to continue through the first couple of years, something to ease the dislocation. In searching for a topic, I contacted Bob Cook, a biologist at the Cape Cod National Seashore. He had a graduate student working with Fowler's toads and was familiar with the publications from my dissertation. After hours of discussion in his office and while sharing a ride to the Cape Cod Natural History Conference, Bob suggested studying the movement patterns of box turtles. It was something he wanted done. With this data, we could determine the home range size and, most importantly, how far the turtles traveled to their preferred nesting sites. He could get me started and provide the equipment. I knew nothing about box turtles other than the fact that they are terrestrial, but this seemed like an interesting project that would fit the bill.

My first day out, a sunny May 20, 2011, was in conjunction with a state-wide survey of box turtles conducted by Lori Erb of the Massachusetts Division of Fisheries and Wildlife. Bob Cook, Lori, her assistant, and I each walked predetermined paths in the woods along the edge of a power line right-of-way. It was the time of year when the abundant pitch pines release their pollen. Unlike many plants that attract insects to carry their pollen, pine trees rely on the wind to do the job. Most pine pollen does not reach the intended target, so to make up for this enormous wastage, pines produce an enormous quantity. A puff of wind at the right time generates yellow smoke among the branches. Our clothes were yellow. My nose began to run, and when I blew into a tissue, it turned yellow. There was pollen everywhere all day long.

On that day, we found four turtles. Each was weighed, measured for length and width, and photographed. Notches were filed into the edges of the shell in a unique pattern, providing a permanent identification for each turtle. Then, a radio transmitter was attached to its shell with epoxy glue. Over the summer, I similarly processed every new turtle I found. Each turtle's transmitter had a unique frequency, so when I wanted to locate a specific turtle, I would dial in its frequency and listen for the pings in a set of earphones. So far, so good. Then the fun really began.

Half-grown Eastern box turtle with epoxied radio transmitter and identifying notches in edge of carapace. Photo by author.

I had seen many nature programs where a biologist followed a wolf, a lion, or some other large charismatic animal by simply rotating their hand-held antenna and moving in the direction of the loudest pings. It looked so easy—not so in the forested sand hills of Cape Cod. I would frequently follow the loudest pings and end up in a location where the pings became diffuse and did not reach their intensity when an animal was nearby. Then, the loudest pings would come from a different direction. Radio waves were being reflected among the hills and giving me all kinds of false signals. How frustrating! I could spend forty-five minutes zeroing in on a signal only to find nothing there. As time went on, I was able to avoid these traps, and I eventually became very skilled at the technique. One improvement was in my ability to recognize when the strongest signal was coming from a well-defined point rather than a broader area. To this day, I can't articulate all the other more subtle ways I improved. This kind of searching is an art, not the precise science it appears to be to the uninitiated.

Each time I located a turtle, there was more than the triumphant feeling of reaching the pot of gold under the rainbow. It was not just the turtle's presence but what it was doing that held my interest. A couple of times, I found my radio-tagged males copulating with an untagged female. A few times in late August, the radio pings led me to an apparently empty stream bottom, but when I moved the dead leaves on the stream bed, there it was, the distinctive curve and characteristic brown and yellow pattern of the turtle's shell. Box turtles like to have a long soak in the hot, dry days of late summer. In the fall, I would find turtles buried in the forest floor, preparing for winter. I was impressed with one turtle that, after spending the summer over a quarter mile away from its wintering site, buried itself about two feet away from that site the following fall.

I always happily did fieldwork alone or with a like-minded partner in my own quiet world. I frequently felt self-conscious interacting with "the public." I never doubted the value or appropriateness of my work, but facing doubters from the "Real World" was usually an uncomfortable experience. Cape Cod was the exception. Following turtles wherever they wandered meant sometimes I was tramping into peoples' yards. Everyone I encountered was supportive and often told me about how they helped turtles across roads. (You must always place them on the side to which they are headed.) Most homeowners on The Cape are environmentally conscious, and box turtles are charismatic animals. People think they're attractive and want to look out for them, so the project did not seem as meaningless as my others might have. Also, because box turtle populations are in danger, there was a real-world aspect to the study.

Having to go where the turtles were, rather than choosing my path through the woods of Wellfleet, frequently meant encountering greenbrier. It was often hot, around 85 degrees Fahrenheit, and I had to wear heavy jeans and hiking boots. I could hardly move as I carefully picked my way through the greenbrier's razor-sharp spines that penetrate clothing and cut skin. The plant is also an entangling vine that forms impenetrable thickets, but when a turtle was under there, I had to find my way, holding some vines aside while I bent down and gingerly maneuvered through. Try as I might, I occasionally got punctured and pissed, and I was reduced to enduring unvented frustration as I sucked the blood running down my arm.

Greenbriar, Wellfleet, Massachusetts. Photo by author.

Not only was the work sometimes painful, it was potentially deadly. On one occasion, I was tracking a turtle and came upon a little stream, part of the braided network that the Herring River forms as it flows through a flat swampy valley bottom. I stood on the muddy edge of the water, slowly swinging the antenna and getting a very strong signal from across the stream. There was no obvious way to cross at that point. I looked around and found a small dead tree lying on the ground. I stood it up on the bank and let it fall across the stream, creating a bridge. Clutching all my equipment, I started inching along the tree, tiny step by tiny step, working to maintain my balance. I was almost all the way across when the tree snapped, dumping me into the water. The water was less than a foot deep, and I remained upright, but the bottom was soft mud, and I sank up to my knees. The more I struggled, the more I sank. I stopped moving to assess the situation. This was serious.

I was aware of the liquefaction of sediments when they are agitated, the same property that makes mortar so effective. The more you struggle, the softer it gets, and the faster you sink. I searched the near shore for a handhold. By leaning over, I was able to grasp the branch of a well-rooted shrub and pull myself out. I don't know how deep the mud was, but it represented the accumulation of thousands of years and could easily have been much deeper than my height. It ended well, but it made me think that there's a risk when working deep in the forest alone. There was no cell phone coverage there. Had I sunk deeper, I might have been unable to extract myself. I had not told anybody precisely where I was working that day, and a search could have taken weeks to find me. What a sobering experience! I never tried to cross a stream in that swampy section of my study area again. Data is important, but not that important.

Nature, Nurture, and the Need to Communicate

One of the facets of teaching General Biology that really got me going was the levels of organization from the molecular to the ecosystem. I enjoyed integrating them and stressing that we break biology into levels for convenience, but it is all one thing. A good example is the idea that natural selection operates on genes. Most models of evolution describe the changing frequency of genes in populations. However, genes do not exist out there naked, waiting to be selected or eliminated. They exist in the cells of complete organisms, and it is the whole organism with its tens of thousands of genes that is successful or not. The genes, in concert with one another and under the influence of the environment, dictate the precise characteristics of the organism that develops. It is these properties interacting with the environment that determine how many offspring are produced, which is the measure of the success of the organism and its component genes.

My views on human behavior are rooted in my understanding of the roles of genes and environment in the development of the individual and the role of evolution in generating the suite of genes in a population. I have had many satisfying discussions with science colleagues about these issues. I also have been taken aback by the response of many of my non-science colleagues to the same discussions. They ranged from polite disagreement to outright anger. This was my personal introduction to the nature/nurture debate.

In the 1970s, there was a resurgence of evolutionary thinking about human behavior, and of course, that meant a role for genes. To many, that implied a return to the horrible eugenics movement of the 1930s. Anyone who entertained the idea of a genetic component to human behavior was considered a racist, sexist, and classist boor. I never held those beliefs, but in the 70s, when Myra, my Hydrolab colleague, expressed confidence in a strong genetic influence on our behavior, I took exception to the idea. I had no reason to prove her wrong, but my approach was that in the absence of a solid data-based argument, I preferred the working hypothesis that we are born as blank slates to be written on by our experiences. I have now seen the data and strongly believe in the genetic contribution to our behavior. I also have had two children and am impressed, as are many parents, with the early emergence of life-long personality traits. From the beginning, Serena was demanding and very much wanted to be in control, while Justin was more passive, even compliant. E.O. Wilson comprehensively made the case for a genetic component to our behavior in the controversial book called *Sociobiology: The New Synthesis*.[59]

The idea that genes influence our behavior is a subtle one. It is no longer "nature *versus* nurture" but rather "nature *and* nurture." We need genes to code for traits, but traits can form only with environmental input. The big question is how responsive is the state of the trait to that variable environmental input. For example, we develop five fingers pretty consistently, but while we are programmed to learn a language, the specific language we speak is completely dependent on the environment in which we grow up. The development of most behavioral traits falls in between these two extremes.

When biologists talk, they often use short-hand terms to express complex ideas. They may say there is a gene for musical ability. What they really mean is that there is a suite of genes of varying influence that contribute to musical ability when exposed to a suitable environment. The corollary is that two people exposed to the same environment could develop very different musical abilities. That's cumbersome in a conversation, so we say there is a "gene for." But if you do not share a detailed knowledge of developmental patterns, you may think that the speaker is saying genes are destiny. And that is the root of the problem. Things can get dicey when a word has different connotations for a speaker and a listener—words can generate misunderstanding, and misunderstanding can foster animosity.

During discussions, when opponents of an evolutionary perspective engage with biologists, their opposition makes it very difficult for them to patiently listen and absorb the nuanced mechanism. The result is an unfair portrayal of such biologists as modern eugenicists. Perhaps the most extreme example of the emotional reaction such thinking engenders occurred in 1978. E.O. Wilson was presenting a lecture at a meeting of the American Association for the Advancement of Science when an audience member came forward and poured a pitcher of water over his head.

Certainly, sociobiologists do not always present ideas that are well supported by evidence. Sometimes, they make pronouncements that are largely conjecture. This tendency is what a prominent Wilson opponent, Stephen Jay Gould, called "just so stories." A hypothesis may sound reasonable in relation to an established theory, but it must not be accepted without corroborating evidence. Nevertheless, it remains true that many good sociobiological ideas are rejected because many people do not want the world to be that way, and they don't take the time to truly understand what is being said. The first step to working out the argument is to understand words in the same way.

The way to achieve mutual understanding is to break out of our bubbles. Many scientists get into trouble when they spend all their time talking with other scientists who share their assumptions and information base. If they try to address non-scientists, they use the same language as they do with each other and, of course, are misunderstood. When non-scientists talk with scientists, they do not understand the scientists' assumptions and information base, so they misinterpret them. Randy Olson has made great strides in solving this problem. Among his many contributions is his aptly named book *Don't Be Such a Scientist*.[60]

In addition to the nature-nurture controversy, this communication problem has been central to the "debates" surrounding evolution, climate change, and many other science policy issues. Our lives are not impacted when academics have this fight, but when legislators don't understand science, we end up with harmful policies that affect everything from the food we eat, the cars we drive, the education we give our children, and the way we treat each other. Thus, the communication problem is not just a rarefied academic issue.

Nurturing Nature

In the early 1970s, a friend gave me a sourdough starter, and I had a new activity involving the tactile and aromatic world of microbes. This microbe connection was not new—my mixing of potting soil for houseplants had also involved manipulating the environment of bacteria and fungi. The sourdough came with a many-time-photocopied one-page description of its ancestry, purportedly having been maintained continuously since its origin during the Klondike gold rush in the 1890s. I have no idea if that was true, but it was a powerful culture. When mixed with flour and a little honey, big bubbles would appear. It had the vigor to raise pure rye flour, something that everyday baker's yeast cannot do. Every Saturday I took the starter out of the fridge and baked bread or made light, fluffy pancakes, always saving a little of the batter in a jar to put back in the fridge until next week. I especially enjoyed the bread making. After mixing the ingredients, I left the dough in a covered bowl to rise. With normal yeast the dough doubles in volume, but this sourdough tripled in volume. The fun part then started—punching down the dough, folding it, and kneading it. The wonderful aroma of bacteria and yeast filled the kitchen. After a couple of years of nurture, I passed the sourdough culture to a friend to keep going while I spent a sabbatical year on St. Croix. To my great disappointment, on my return I learned that she had left too much time between baking days, and the culture died.

Unlike fragile sourdough microbes, most soil bacteria and fungi are quite robust. In contrast to the uniformity of bread dough, they live in a highly structured, dynamic system—soil is a layer cake of distinct levels grading from very organic at the surface to very mineral at depth. Different bacterial, fungal, and animal inhabitants occupy each layer, and they are very abundant. A handful of average soil contains billions of organisms. With particular physical and biological components, the chemical nature of each layer is unique. If you dig a hole, keeping the sides smooth, the layers will be apparent by their different colors and sharp boundaries. In other words, "dirt" simply does not cut it as a descriptor.

Gardeners crave loam, a light soil with a favorable mix of sand, silt, clay, and organic matter. Loam is composed of small lumps, giving it abundant pore spaces that hold water but also let air penetrate so roots can breathe. I always kept potted plants around the house, and one of my more pleasant chores was mixing potting soil. This usually started with a couple of shovelfuls of soil from the garden, which was often somewhat heavy. I then added some sand to make it more porous and compost to add nutrients and lighten it up. To have a better feel, I mixed it with my bare hands, breaking up the lumps and creating a uniform, dark soil that held water well but did not get waterlogged. During the mixing, I took pleasure in the earthy smell released by the various bacteria and fungi residing there. That smell told me that I was creating a healthy, living substrate that would foster vigorous plant growth.

Bread is not soil, and soil is not bread, but they have served me in similar ways. Both provided a stress-relieving tactile experience as I mixed them. Both released a pleasant organic aroma that had a calming effect on me. And both gave me the pleasure of providing a nourishing medium for the growth of beneficial microbes.

On the Otter Hand

In 2009, when we purchased our house in Truro, Cape Cod, it came with a little pond, six feet wide by nine feet long. I was not looking to manage a pond, but once committed, I enjoyed taking care of it. There was a leak in the liner, and during the first winter, because it was only half full of water, the pond froze solid, and I was greeted at the spring thaw with two dead frogs floating on the surface. I purchased some plants and ten "feeder" goldfish at fifteen cents each. I frequently added water to keep up with the slow leak. The pond looked pretty good all summer, but the water was green.

The following spring, I removed the old liner, dug out a hole in one spot to increase the water depth as insurance against a solid freeze, and installed a new liner. I added a few more plants along with a UV filter to kill phytoplankton and prevent green water. This was a major compromise on my part. In the various aquaria I have maintained over the years, ranging from simple fish tanks to an enclosed coral reef, I have always worked towards achieving an ecologically balanced system with minimal technology. If you have the right mix of species and don't overstock the system, it should be self-sufficient. That means accepting algae and bacteria as natural components that keep the water quality suitable to support the fish and other fauna. This pond was different, however. I wanted goldfish to add interest for my guests and grandkids. The fish food added nutrients to the water, and along with the intense sunlight, those nutrients allowed phytoplankton to thrive. The goldfish were lost in the pea soup. The UV filter selectively killed the phytoplankton, which allowed light to penetrate the water. Consequently, thread-like algae covered the liner, providing a different ecosystem component to absorb the nutrients in the water. The goldfish thrived, developing hefty bodies and producing new little goldfish each spring.

I had a pretty pond with clear water and interesting fish for the non-ecologists to admire. This system worked beautifully for several years. A couple of times, a fish disappeared, and I attributed this to the visit of an egret or heron, though I never saw one there. One day in early summer 2015, the pond looked as though someone had taken a giant mixer to it: the potted plants were turned over, there was debris and floating plant pieces everywhere, and a goldfish head lay on a rock at the edge. What happened? The next morning, I got up early and stealthily approached the pond. The water was roiling, and a sleek, brown back was arching at the surface and resubmerging. Then a head appeared, the head of an otter. The otter was circling rapidly in the tiny space as it chased my prized goldfish. I knew that simply chasing it away would result in its return in the future. I decided to give it a strong negative experience so that it would associate the pond with fear rather than food.

Goldfish in our pond before visit by otter. Photo by author.

So, I went to the garage and grabbed a five-foot-long fiberglass plant stake that had a pointed end. Sneaking back to the pond, I saw the otter continuing its tight, confined circles. I really didn't have to sneak because it was so intent on fishing that it was unaware of me. I poked the otter a few times, and it began swimming to escape rather than to eat, but it stayed in the pond. I persisted, and at one point as the otter passed the far edge of the pond, I jabbed at it with the stake. In rapid succession, I felt resistance as the stake pressed against the otter's belly and then a sudden release. I must have punctured its skin. It was a sickening feeling. The otter burst out of the pond and sprinted off in a panic. I had intended to scare it, not harm it. Would it heal and be okay, or get infected and die a slow, painful death? I didn't know, but the thought stayed with me for days and days afterward.

I knew there were otters in the area. Brigid came face-to-face with one while swimming in the upper reaches of the Pamet River. From her perspective and in the shock of the moment, she fearfully wondered what an anaconda was doing there, global warming? Otters appeared in Village Pond a few years earlier and ate out all the carp, resulting in that pond becoming choked with water milfoil, an invasive rapidly-growing aquatic plant. My otter may have come from there or somewhere else that was exhausted of fish. Or, it may have been a juvenile that was forced out by its parents as a new generation of baby otters arrived. In any case, in spite of my guilt, I felt I had at least solved my otter problem. Several large goldfish survived, and after some cleaning up, the pond returned to normal.

But one day later that summer, I saw that the pond had again been blendered. I checked the following morning, and it had happened again. I could not see any goldfish remaining. Had that otter returned? Was the memory of being pierced overtaken by great hunger? Or was it just a different otter? I didn't know. After the ice melted in the spring of 2016, I could see tiny gold splotches moving about the bottom of the pond. The next generation of goldfish was preparing to take its place. I wondered: *did the otters know the pond had been emptied of eating-sized fish, or did they see it as a new source of food?*

Feeding the goldfish was a treat for our grandchildren, so we wanted reasonably large fish in the pond and purchased three more. Soon afterward, the pond was disturbed again, and the fish were spooked, hiding under the lily pads. Instead of leisurely coming up to ingest the floating food pellets, they would wait a while and make a mad dash before quickly returning to their shelter. I soon noticed that the newly purchased larger fish were gone—and some of the smaller ones, too. Because of the reduced blender effect, I wondered if a raccoon or some other predator was the culprit until a man renting the house came up to me and said: "Do you know you have otters in the pond?"

He showed me an iPhone photo he had taken that morning. There wasn't AN otter; there was a mother and two juveniles—three otters in that little pond! It was a perfect setup for hunting "training wheels!" Was the mother the otter I had speared the previous summer? I had no idea.

The following April, once again, there was an otter in the pond, and I took some pictures of it. I downloaded the shots and stored them on my computer with all my other photos. A year and a half later, I was reviewing my photo collection and came across the otter shots. Although I had not noticed it originally, I now saw that the otter had a long white scar on its left side—the same side that I had pierced on that day three years earlier. I finally had my answer. The otter survived my attack and lived to make more trouble.

When people complain of deer, woodchucks, and other wildlife in their gardens, I've always said the animals were here first. We moved into their habitat, and they have a right to be there. I still believe that. So, I now had a pond with a few small, traumatized goldfish that my grandchildren did not feed. Really, it's a small price to pay to live in a wonderful, healthy environment full of wildlife. Actually, the price was slightly higher. In the spring of 2020, when I did the annual cleanout of debris, I was struck that no frogs had overwintered in the pond. Otters eat frogs, too. One day, as I sat in the house reading, I saw an otter run across our back deck. I quickly went to the front window to check out the pond. I saw the otter make two quick underwater circles and run off. Now, I knew for sure that we were on her circuit and would never have goldfish while she lived. River otters live fifteen to twenty years in captivity but only eight to nine years in the wild. So, there's hope that I might return to my previously populated pond, and the only way I will know to do that is to introduce new goldfish until they persist. But I'm not about to do that for a while.

Using Scavengers

Aside from pond maintenance, I enjoyed gardening, photography, and writing. I also had a bright idea. I would prepare a fish skeleton as a piece of art to hang on the wall. Among other things, it would be a way to explore something that had intrigued me for a long time—the use of beetle larvae as a way to clean the flesh off bones. Many museums keep dermestid colonies specifically for that purpose. I don't know how many still do, though, now that models and molecular techniques are replacing mounted skeletons for both research and display.

First, I had to get my fish, and I wanted one from the sea. I have fished in the past, mostly with my father in freshwater lakes, but I'm not a skilled saltwater fisherman. The obvious solution was to get a fish frame from a fish market. A frame is the skeleton that's left after the fillets have been removed. Our local market in Truro, Cape Tip Seafood, receives their fish already filleted from their wholesale shop in Provincetown, so I called there. They were willing to provide me with a frame. After all, it was waste material to them. After a few days and my making a pest of myself with the frequent calls, they finally said they had a striped bass frame for me. When I went to pick it up, I was presented with a three-foot fish skeleton packed in ice—an extra touch I had not expected. Obviously, I should have tipped the guys for their effort but somehow I didn't and felt pretty guilty afterwards. A lot of good that did—the transgression remained.

On getting the fish home, I set it on my workbench and began to remove as much flesh as I could using a scalpel. The flesh was cool and smelled like the ocean. I tasted a piece. It was firm, delicate, and complex, so I removed as much as I could in bite-sized pieces, collecting them in a bowl sitting on ice. Brigid and I got out the soy sauce and a little wasabi and enjoyed a wonderful sashimi dinner. That missed tip to the fish guys was now even more egregious.

It was a delicious start, but this project was about the bones. A few weeks before picking up the fish, I went online and looked for a source of dermestid beetles. I found several and discovered in the process that they are used primarily by taxidermists to clean the flesh off the bones of trophy specimens. I ordered some, and in a few days, they arrived at the post office. I placed them on a bed of wood shavings in a plastic container and fed them pieces of chicken. They were happy and started making babies. Baby dermestid beetles are white, worm-like larvae that squirm around and gorge on flesh. It's not a pretty sight to most people. To me, however, it was a job being efficiently performed, and I found that appealing.

Once the skeleton was as clean as I could get it, I placed the fish on a bed of wood shavings in a large plastic storage container and added the beetles. Because we have raccoons, coyotes, and other scavengers in our neighborhood, I placed the container in our garage—big mistake. In the summer, we rented out our house and stayed in an apartment over the garage. I did not have a large enough beetle population to deflesh the bones in a few days, the usual timing and the flesh began to rot. One day, Brigid asked: "What's that awful smell?" and I was forced to admit my miscalculation. The solution was to take the garbage cans out of our wooden trash bin under the stairs and put the fish in its container in there. Now, the project could be outdoors and still protected from scavengers. Relocating the fish eliminated the smell coming from the garage, but we still got frequent whiffs of stench coming in the window on the gentle breeze.

As time went on, the beetle population grew, and there was an impressive number of larvae and adults scurrying about the bones. It took a couple of weeks for the beetle larvae to do their work. Fresh or rotten, it was all the same to them. I had expected the ligaments to persist, holding the bones together, but they were consumed, too. So, instead of an intact fish skeleton, I had a heap of loose bones. That ended up being okay because I had not appreciated the oiliness of fish bones. I found out from those taxidermy sites that the solution is to soak the bones in warm water and dish soap. I kept the water warm by using an aquarium heater I had lying around.

Completed striped bass skeleton. Photo by author.

It took several soaks of two days each, and after several rinses and spreading the bones out in the sun, I ended up with a pile of clean, dry, odorless fish bones.

Now, the assembly could begin. I had taken photos of the skeleton but could not identify several skull bones. The skull bones of fish are more loosely attached than those of mammals, and these had completely separated. With the help of photos and drawings of fish skeletons on the internet, I was able to distinguish the bones and fit them together like a three-dimensional jigsaw puzzle. The tail fin consisted of thirty separate bony fin rays whose elements thankfully remained intact, and I used 100 pins to hold them in place while they dried. I glued all the vertebrae, the skull, and the tail onto a heavy aluminum wire, one end of which I had fashioned into a loop for hanging. This attachment was a tedious process because I had to wait overnight for the glue to harden on each bone before connecting the next. Eventually I was rewarded with a completely assembled skeleton, very much the way I had envisioned beforehand. I was ready to display my art, but the bones were white, and the wall on which they were to hang was also white. I needed to create some contrast. I struggled with colors in my head but decided there was only one that would do. I spray-painted the bones with black enamel, and there it was, nicely contrasted against white surfaces—a big head connected to a long backbone ending in a fan-like set of tail bones three feet from the head. It turned out that black, the color of death and mourning, was a good choice for the impact of the final piece.

Evolution and My Mind

I live in a Newtonian world. What does that mean? Isaac Newton gave us the basics of classical physics: the laws of forces acting on matter. Classical physics is very different from modern physics, popularly characterized by Albert Einstein and $e = mc^2$. Newton tells us that the application of energy in the form of force can move matter, whereas Einstein tells us that energy and matter can be converted from one to the other—they are different manifestations of the same thing. Modern physics explores what happens near the speed of light or near absolute zero (the complete absence of heat). We even hear of subatomic particles separated by miles interacting with each other or that we exist in one of innumerable parallel universes. I have trouble distinguishing some of modern physics from philosophy.

My mind is the product of evolution. It was molded by millions of years of natural selection in which protohumans existed and functioned on a time scale of seconds to decades and a spatial scale of fractions of an inch to miles. While my brain is flexible and can conceive of an immense range of concepts, it is also bounded by built-in limitations. It is not a blank-slate computer that can be programmed to do anything. It is a dedicated computer, prewired to acquire language, detect cheats, and think about a physical world on a time and spatial scale in which a foraging hominid exists on a savanna.

We can calculate all kinds of things that we cannot conceive of. For example, the number of molecules in about a half-ounce of water is 602,200,000,000,000,000,000,000 (that's 602 quintillion). David Brewer, my chemistry colleague at Sarah Lawrence, once told me that if a molecule were the size of a ping pong ball, this number of balls would cover the entire US to a depth of nine inches. Does this help? I cannot conceive of this because I cannot conceive of the 3,618,780 square-mile area that represents the continental United States, even though I have flown over it many times. We use all kinds of tricks to help us overcome our limited brains and visualize big and small numbers. They help very little. For example, when the US national debt was $17 trillion in the summer of 2014, Franklin Thomas

Revere, an information technologist, calculated that those dollars laid end-to-end would circle the Earth 64,000 times. Does that help you comprehend that number? Research with animals shows that many species can count up to three or five or some other limited number. Although our ability to use symbols allows us to write down and manipulate extravagantly large numbers, we, like other animals, have a conceptual limit to what we understand. There is a number, call it "very big," beyond which we cannot fathom.

So, while modern physics explores strange phenomena on the very small subatomic scale and the very large interstellar scale, my very powerful but very limited hunter-gatherer mind cannot conceive of these things. Being a field biologist, the classical physics of Newton provides explanations for all the phenomena that I encounter, from how a fish swims to the atmospheric processes determining climate. While molecular techniques and modern computers are tools that I use as a twenty-first-century field biologist, I am still working in a world ruled by seventeenth-century Newtonian physics.

The disconnect between what we can convey with numbers and what we understand in our guts is evident in many other ways. One of these ways keeps us scientists as caring, loving, compassionate human beings while we manage to varying degrees to look at the world, including ourselves, as objective observers. Try as we might in our professional lives, and despite all the portrayals in the media, scientists cannot be like *Star Trek*'s Mr. Spock.

I have experienced this disconnect in many ways, but probably the most striking is in my understanding and feeling about living things and how they function. I do not spend much time thinking philosophical thoughts, but I can clearly state that I don't believe in the supernatural in any of its forms, including religion. I am comfortable living in a material world in which all phenomena are ultimately explicable by the "Laws of Nature," which we still don't fully understand. I comprehend the theory of evolution, a simple yet often misunderstood idea. Its modern application is so intellectually satisfying, especially in the realm of "evo-devo." Evo-devo looks at embryological development in terms of the sequential turning on and off of genes, showing how body patterns emerge as the fertilized egg develops. It also shows how the timing can change and how the genes can change, resulting in different patterns and, thus, different species. We are getting to the point where we can see how individual genes impact the ability of the individual organism to survive and reproduce. There is much more to learn, and all this only makes sense in the context of the natural environment of the organism, but we have seen remarkable connections and can envision those to come. Even though I do not know the precise steps by which any species has evolved, I know enough to be confident that every living thing on Earth has descended from a single life form that arose over 3.5 billion years ago. The mechanism, if not the details, is known.

And yet, it's utterly astounding. The diversity is so immense that for all my years of study, I can still watch any nature program on TV and almost always learn something new. For all my intellectual understanding, I am awed by the remarkable precision and unexpected complexity of adaptations from the molecular to the behavioral. Many such examples have been described by authors of books on evolution. If you want to explore one, read Chapter 2 of Richard Dawkins' *The Blind Watchmaker*,[61] in which he describes the remarkable capacity of bats to track flying insects through the use of ultrasound. He states that ". . . their 'radar' achieves feats of detection and navigation that would strike an engineer dumb with admiration." Bear in mind that for all the surprising sophistication he recounts, we have since learned that the details are even more nuanced than we knew when he wrote that book.[62]

Some people say that one characteristic of the human mind is that it can hold two opposing ideas at the same time. We are all capable of logical inconsistencies, but maybe we don't believe these things at exactly the same time. Perhaps it's more like that optical illusion, the drawing that looks like a vase or two faces. We can see one or the other but cannot perceive both simultaneously. Similarly, I am overwhelmed by the unimaginable intricacies of living things. I have an emotional response—of a sense of beauty, of precise coordination, of amazing perception, of raw power, of parental love, of repellent decay. For example, I'm blown away at the thought that some birds have their migratory route coded in their genes. How in the world do sequences of T's and A's and G's and C's code for direction, distance, and even a map sense based on star patterns or the Earth's magnetic field? On the other hand, I understand random variation, differential survival and reproduction, and incremental change over millions of years. I guess this is a reflection of the old "left brain-right brain" dichotomy. I absorb things in different ways that cannot be resolved, and I wouldn't have it any other way.

Evolution Isn't Always Efficient

Rube Goldberg was known for producing cartoons depicting complicated gadgets that perform simple tasks in the most indirect, convoluted ways. His cartoon for a simplified pencil sharpener is a good example:

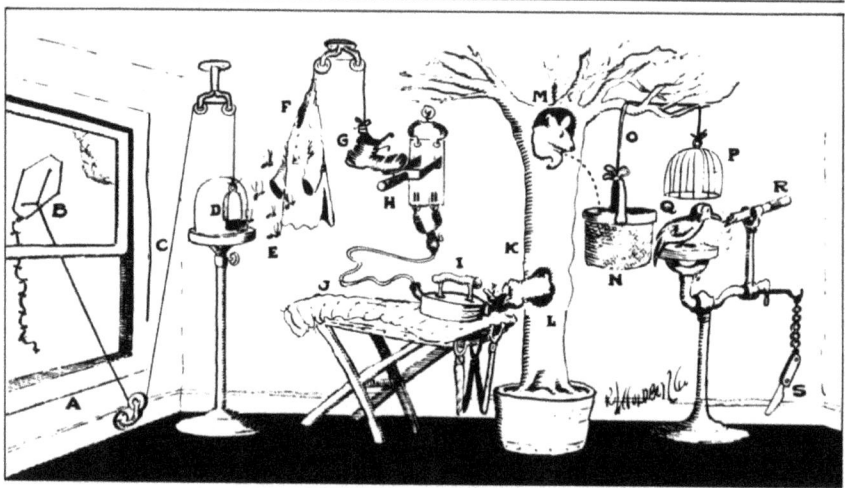

The Professor gets his think-tank working and evolves the simplified pencil sharpener.

Open window (**A**) and fly kite (**B**). String (**C**) lifts small door (**D**), allowing moths (**E**) to escape and eat red flannel shirt (**F**). As weight of shirt becomes less, shoe (**G**) steps on switch (**H**) which heats electric iron (**I**) and burns hole in pants (**J**). Smoke (**K**) enters hole in tree (**L**), smoking out opossum (**M**) which jumps into basket (**N**), pulling rope (**O**) and lifting cage (**P**), allowing woodpecker (**Q**) to chew wood from pencil (**R**), exposing lead. Emergency knife (**S**) is always handy in case opossum or the woodpecker gets sick and can't work.

Artwork Copyright © Rube Goldberg Inc. All Rights Reserved. RUBE GOLDBERG ® is a registered trademark of Rube Goldberg Inc. All materials used with permission. www.rubegoldberg.com

We usually think of evolution as a mechanism for producing the most efficient, least wasteful structures. Biology books and TV nature programs are filled with excellent examples of beautifully synchronized behaviors and physiologies that maximize evolutionary fitness. But some situations prevent evolution from achieving perfection. Sometimes, the result of evolution is an awkward workaround that does the trick but is not the elegant solution we have come to expect. To illustrate this point, let me contrast the divergent solutions achieved by birds and mammals to the evolution of high, stable body temperatures.

The testes of birds are tucked deep in their body cavities, where the testes of almost all organisms reside. Why do I make this point? Because the testes of mammals reside outside the body cavity in thin-walled sacs. Imagine the mammalian solution applied to birds—sparrows flying about with little testicles dangling below them. Why this difference? Virtually all biochemical reactions in the bodies of living things are regulated by enzymes, which are proteins that function as very specific catalysts. Many enzymes function optimally in a narrow range of temperatures. When birds slowly evolved warm-bloodedness as they diverged from their reptilian ancestors, presumably genetic variations were selected that resulted in increased optimal temperatures for hundreds of enzymes. In this way, the biochemistry underpinning all functions continued to operate in a harmonious, coordinated manner, and all organs remained in their historical locations.

In contrast, mammals relocated their testes. What is arguably the most valuable organ of male mammals sits in a vulnerable, unprotected position. How odd that an ancestral internal organ is suspended outside the body in a thin-walled sac. During ejaculation, sperm has to travel through tubes from outside the body back into the body cavity, where it encounters various glands producing semen. Then the mixture travels back out of the body cavity to the penis. Why would evolution produce a Rube Goldberg solution to a problem when a clean, simple solution is possible?

Natural selection can only choose between existing genetic options that arise randomly. If, by chance, mutation does not provide the genes coding for the optimal solution, then the next best solution will emerge. What probably happened (and this is another "just so story") is that a mutation coding for the higher temperature that is optimum for the enzymes involved in sperm production did not appear. However, other genes affecting the location of the testes did appear. Slowly, over generations, as the body temperatures of early mammals increased, the testes moved closer to the body wall, where it was a little cooler, and they eventually moved through the body wall into a slowly developing scrotum. In this way, the testes were able to have a lower temperature than the body core, and thus, their enzymes continued to function effectively. So far, so good.

But what are the drawbacks to such a solution? We all know the easiest way to make a threatening man double over on the ground. And for some reason, videos of men being hit in the groin by rakes, balls, and various other objects are featured as hilarious clips on "funny" home video TV shows. Sometimes, style affects our virility—tight pants cause warmer testes and reduced fertility. Beyond that, the descent of testes through the body wall results in an opening, the inguinal canal, that allows nerves, blood vessels, and spermatic ducts to pass through. This channel constitutes another weakness, and combined with the high pressure in our abdominal cavity due to our upright posture, creates a problem: a loop of the intestine can slip into and through the canal (a hernia). If the intestine twists, the blood supply can be cut off, and tissue death can occur. I have had hernia repairs on both sides. Now I'm protected from that threat. Eyeglasses, hearing aids, appendectomies, knee replacements, and many other interventions are the many ways modern medicine covers for evolutionary shortcomings.

So, bipedalism has created a problem not only for women in childbirth. Men have their issues, too. The optimal solution achieved by birds is clearly superior to the cumbersome solution of we mammals. As long as the benefits exceed the costs, evolution will move an organism in a given direction. It may not take us to an ideal place, though, just one that is better than where we were.

This "good enough" solution increases our reproductive fitness. But what happens when reproduction ceases? I'm retired, out to pasture, but I can still contribute a lot to the people around me. I write, take photos, give talks, maintain my house, donate to good causes, and consult about local environmental issues. But biologically, I'm useless. By that, I mean I have nothing to contribute to the perpetuation of my genes. I am excess baggage, taking up space and resources but giving nothing back. If I died today, my grandchildren would do just fine.

The Meaning of Life, A Meaningful Life

Nature has no morals. There is no good and bad, only relative degrees of success, measured by the number of copies of one's genes contributed to future generations. It has to be this way because it is how we came to be, starting with self-replicating molecules over four billion years ago. Since then, there has been continuous shuffling, testing, and selecting, resulting in the wonderfully rich diversity of life as we know it today. Everything from the simplest virus to a human being was produced the same way and, therefore, behaves the same way; it makes copies of itself. From a purely biological perspective, this is the only meaning of life. Instead of saying that "an egg is a chicken's way of making another chicken," we can say that "a chicken is an egg's way of making another egg." By reversing the statement, we focus attention on the genetic material in the egg, and "chicken" can be replaced by "daisy," "honeybee," or "person." The bodies of all organisms are simply devices to maximize the perpetuation of genetic material. Nothing more.

So, am I a depressed nihilist? Not at all. I love the people around me. I love life. I'm programmed to be this way. Knowing that I am living out a scenario coded in my genes does not lessen the feelings I have. I intensely love my children, and I like to look at naked women. I have been conditioned by society to celebrate my love for my children but not to admit to looking at naked women. Our genes predispose us to be social beings, making us subject to the mores of the society in which we live.

Our large brains have defensive mechanisms that allow us to function as designed. One way this happens is by selective attention. I know that I am the product of a mindless selective process and that my life has absolutely no meaning outside of what I give it. This knowledge usually resides below my consciousness and has no impact on my daily activities. Even when I focus on those facts, I can simultaneously experience all the warm feelings I have for the people around me. The cognitive brain does not trump the emotional brain—a connected, engaged brain results in evolutionary success.

In this way, the brain has built-in reinforcing mechanisms to ensure reproductive success. We avoid pain because we cannot reproduce unless we survive long enough, and we seek sexual pleasure because survival without reproduction is evolutionary extinction. The body has repair mechanisms to keep it operating effectively while reproductively active. These repair mechanisms have been selected to operate while they impact the success of my offspring. Being human, my offspring needed many years of care while they were developing, but once they became independent, I was superfluous, and selection has not operated to maintain my body beyond its reproductive usefulness. So, now I have memory problems, arthritis in my knees, an enlarged prostate, and the associated aches and pains that come during this period of slow decline.

As far as we know, the human animal is unique in seeking meaning in the world around it. Evolutionists believe that this trait is an adaptation to solving problems and thus improving the chance of survival. I believe this attribute is the root of all art and science. We are unfulfilled if divorced from creativity. We marvel at the beauty and control of a pelican soaring just above the water's surface, the tips of its wings skimming the membrane between two worlds. We are enchanted by the ethereal warbling of the wood thrush somewhere among the trees of a forest. We are almost overwhelmed when we inhale the sweet fragrance of lilac blossoms. We run our fingers through the smooth pebbles at the bottom of a shallow mountain brook, feeling the pressure of the current on our hand and the cool, soothing water, and hearing the stones rattle as the water gurgles. We are embedded in the natural world, and we love it. Mary Oliver said it this way: "Nature . . . is the wheel that drives our world; those who ride it willingly might yet catch a glimpse of a dazzling, even a spiritual restfulness, while those who . . . insist that the world must be piloted by man for his own benefit will be dragged around and around all the same, gathering dust but no joy."[63]

Nature is variable. Every day is different. But the variation generally falls in a limited range to which we are accustomed. Every once in a while, though, something falls far enough outside that range and makes a big impression. These unexpected, once-in-a-lifetime moments are different from my encounters with the manta ray and the octopus, however. One such occurrence was otherworldly, the kind of experience that can lead some people to believe in angels or ghosts or embodied spirits.

Early in my career at Sarah Lawrence, on a mid-October morning, I was walking alone on the path up Bates Hill. The empty campus was shrouded in a dense fog, the kind you can almost touch. The tall oaks had already dropped their leaves, their damp trunks black and the spreading bare branches a deep gray, becoming lighter and wispier with height. All sounds were muffled. I felt enveloped in my own cocoon, almost like being in a sensory deprivation chamber. As I trudged upwards, thinking about the events of the upcoming day, a pair of pure white swans emerged from the fog, flying at tree height directly over me. They were completely silent—enormous birds with necks outstretched, one behind and slightly to the right of the other, broad wings slowly beating—ethereal milk-white ghosts against a milky-gray fog. I saw them for perhaps two seconds before they disappeared. They were there, and then they weren't, apparitions in the sky. I stopped, awestruck, looking up at the nothingness left behind. I never saw swans on the campus before or after that moment, but they chose to make their only appearance on that mysterious, magical morning.

You might call me a determinist. You may say there is more to life than simply reproduction. Indeed, there is. The emotions we feel are built in, but their meaning is something we create with our large brains. Art, religion, politics, economics, and all the other cultural features of human societies are made up in our heads. They may or may not increase our biological fitness. I am descending towards the end of my allotted time, and I adapt to my condition, not with enthusiasm but acceptance. I don't look for meaning where there is none, but there are many things I've found in life that are meaningful to me. I am enjoying the life I have and will make the most of my remaining time as an engaged member of the human species.

"There but for the grace of God go I."

No, I have not undergone a change of heart as my end approaches. It's just a useful expression. The point is I have discussed several incidents in my life that could have had tragic results but didn't. I could have died from an infection at the age of seven in the cinder sifter incident. I could have cracked open the skull of a dental student when I swung my stick in anger in the floor hockey incident when I was eighteen. At the age of twenty-seven, I could have run over Karen when the boat circled out of control. I almost burned down Brigid's house in the oily-rags incident at the age of fifty-three. At fifty-eight, I could have died from a Vibrio infection. And I could have perished stuck in the mud of the Herring River at the age of sixty-nine.

Boy, was I lucky to have survived it all. Probably most of you have had similar, equally hazardous situations. That's why we have religion. It gives meaning to these things. Many people find it comforting to think that someone is looking out for them. For me, it's simply my good luck at the rolling of the dice. I'm glad I'm here, but I'm not thankful for some greater power. I am thankful for the scientists who developed antibiotics.

Bringing it Home

Our one-acre property on Cape Cod is covered in almost pure sand deposited by the retreating glaciers twelve thousand years ago. In the subsequent millennia, a layer of topsoil developed, and forest trees moved in. However, deforestation in the eighteenth and nineteenth centuries resulted in erosion and the loss of this topsoil. Patches of blueberry, grasses, pitch pine, winged sumac, and bear oak now cover the site. Only the pines are trees. I have spent much time cutting the winged sumac to prevent its taking over. These bushes have runners, and they can quickly spread. By maintaining this patchy environment, I provide a diversity of habitats that support a range of wildlife, from bumblebees to coyotes.

This mix of vegetation provides little shade, and I felt that a spreading tree would provide relief during the hot midsummer days. I could have planted a Norway maple that would grow quickly, but I wanted to have an indigenous tree. Of all the native trees on the Cape, the white oak is the most majestic. It is also very tough and can withstand the strong winds from the northeasters that come in winter. Its strength is recognized by boatbuilders who value it for frames and planking, but strength comes at a cost. A tree can put its energy into rapid growth by creating low-strength wood. By putting its energy into strong wood, a white oak forfeits rapid growth. I decided to plant a white oak even though it wouldn't reach its glory during my lifetime. This one is for my children and their children.

I called several nurseries in the area, but none carried white oak trees. I decided I would appropriate one from the nearby forest, which had the added benefit of coming from a local population, so it would be genetically adapted to our climate. A nursery tree could have originated in a different part of the country and may have had a different genetic makeup. You might think that taking a tree is a disruptive act, and it is, but the disruption is short-term. As forests grow, there is strong competition among young trees, and many die as others thrive. My removing one tree gave the others nearby a better chance of surviving.

I walked along forest roads looking for a white oak about my height and with a trunk about the thickness of my thumb. I didn't want a larger tree because I would have to cut a bigger proportion of its roots, and that would reduce its chance of surviving the transplant. I found several that looked good until I saw that they had been cut and re-grown. They might have had large root systems. The town had done the cutting in an annual task to keep the roadside vegetation from reducing visibility.

Finally, I found my tree. It was the right size, fairly straight, and close enough to the edge of the road for me to back my car in and lift it out. It was August 2013—a very dry summer. I cut the roots in a circle about a foot from the trunk, and to balance the roots and leaves, I trimmed a couple of branches. Trimming branches also helped shape the tree to my liking. For the next several weeks, I watered the tree. This was my effort to help the tree heal and reduce the shock of being uprooted. I also reduced the shock by waiting until the tree was dormant before digging it up.

I returned with a trowel, shovel, and lopping shears in November after the tree had dropped its leaves. I backed my car through the trees and got to work digging away the soil outside the circle of roots I had cut in August. After digging down about a foot, I began to dig horizontally under the tree. I cut the roots that I encountered, and then, directly under the trunk, I ran into the taproot. Unlike many trees, white oaks have a deep-diving taproot. This root was substantial—its diameter was greater than that of the trunk. I had no choice but to cut it, and I wondered if this was a death knell for the tree. After all, white oaks are notoriously difficult to transplant. All I could do was complete the task and see what resulted. After freeing the underside of the root ball, I worked a piece of burlap across the bottom and folded it over the sides and top, tying the ends to the trunk.

I was now ready to move the wrapped tree, but it was too heavy for me to lift. I put two long boards between the back of my station wagon and the edge of the hole and, with great effort, slid the root ball out of the hole and up the planks into the car. I moved the root ball as far forward as possible, but the tree still extended out the back, so I left the hatch up. I filled the hole with the removed soil and pushed leaf litter over the top, leaving a slight depression in the forest floor. I drove the six miles home very slowly in the break-down lane. A cop car passed me, and I wondered if it would stop and I would be questioned, but it kept going.

Arriving home, I backed the car up to the hole I had dug the day before. It was four feet wide and two feet deep. I mixed the sand from the hole with compost to provide a rich growth medium. This compost was made from the scraps from our kitchen and cuttings from the garden. I left plenty of sand in the mix to provide good drainage and to avoid adapting the tree to unnaturally rich soil. I dug a narrow, deep pit in the center of the hole and filled it with my mixture, hoping to encourage the truncated taproot to regrow. After covering the bottom of the hole with the soil mixture, I eased the tree down the planks. With some adjustments to get the trunk plumb and the base of the trunk at the surrounding soil level, I filled in the rest of the hole with the soil mixture, watering it periodically to ensure the soil settled around the root ball. I mulched the surface with shredded bark, then hammered three stakes in the undisturbed soil at the edge of the hole and tied guys to the tree to provide support against the strong winter winds. Then came the long wait to see if I was successful.

The next spring, I was rewarded with a flush of delicate leaves—so far, so good, but not unexpected. The key was to see if the tree would survive the next two or three years with its damaged root system. To this end, I watered the tree regularly and hammered fertilizer stakes into the soil.

Ten years later, the tree is thriving. It is growing straight and tall with a full canopy. I look at it frequently and have a warm feeling about its healthy appearance. It means so much to me that when I am having trouble falling asleep, I visualize the tree, and it calms me. It's still a sapling with much potential. I'll never know if it becomes a mighty oak, but it will be part of my legacy, a long-lasting sign that I was here and, with this tree, made the world a slightly better place.

Transplanted white oak, first spring on left, tenth spring on right. Photos by author.

A Final Thought

The practice of science is a creative act. New concepts have many different origins, but we are thinking about ideas for the first time in all of human history. This social endeavor is exciting and rewarding. Scientists don't work in isolation; they share information and build on each other's findings while they also critique each other's ideas to eventually come to a consensus. Conferences are places where connections are made and hypotheses freely shared. Being reticent, I did not make the most of these opportunities. Where others plunged forward, leading with their new concepts, I tended to hang back—no crazy head-centered costumes here. My findings may have generated a larger impact had I more forcefully engaged other scientists and marketed my discoveries. Nevertheless, my publications and seminar presentations reached many colleagues and provided part of the foundation for their research. Nothing is as satisfying as seeing my work contribute to the design of other scientists' experiments. Although I am retired now, not producing new papers, my publications continue to be referenced by the next generation of biologists. My work lives on.

I can't think of a more satisfying career for me other than the one I chose. Although my nature predisposed me to become a scientist, working, learning, and teaching in that realm have greatly impacted my mind. I have fully bought into the scientific way of thinking and accept very little at face value. My questioning of every new idea, even in my private life, is a trait that can be annoying to others, and my embrace of evolutionary thought has progressed beyond my science to personal and social issues. I see myself and everyone else as products of natural selection, displaying a biology and behaviors shared with many other animals. I am not unique in this respect. It is an attribute shared with most of my colleagues. But how I got to be this way and how it has informed my life is my story alone.

Epilogue

After I completed this book, Brigid and I had the good fortune to visit the Galapagos Islands, the place where it all began. By "it" I mean the idea of evolution by means of natural selection. I felt that I was travelling in the footsteps of the two men that most influenced me as a biologist. I visited Buccaneer Bay, on the island of Santiago, where Charles Darwin is known to have spent some time, and from a distance, I saw the island of Daphne Major where Peter Grant did the bulk of his fieldwork. Darwin gave me the intellectual framework to understand the living world and Grant turned me on to the role of competition in structuring biological communities. Darwin visited the islands in 1835, Grant first started working there in 1973, and I visited in 2024. But I was only a tourist, poking around and looking for examples of the plants and animals that influenced their thinking.

My primary goal was to observe Darwin's finches. The seventeen species[64] of these birds are used as the iconic example of how the isolation of populations on islands leads to speciation. Darwin did get this insight there, but it was by observing that each island had its own species of mockingbird. The finch story did not emerge until others studied the specimens that he brought back to London. Nearly 150 years later, Peter and Rosemary Grant studied the finches in immense detail and demonstrated that in dry years, fewer seeds were produced and the soft ones, being quickly consumed, resulted in only hard seeds being available. Consequently, finches with the larger more powerful beaks left disproportionately more offspring. The average beak size in the population increased. During wet years, when a large variety of seeds remained available, the average beak size decreased. Even more surprising, the Grants observed the origin of a new species of finch through hybridization in only two generations.[65] These observations illustrated for the first time that natural selection occurs in wild populations on a short timeline, and their work advanced the concept of evolution by means of natural

Green caltrop seed, three-fourths inch from spine tip to spine tip. When mature they are extremely hard and only the large ground finch with its parrot-like beak can open them. Photo by author.

selection from a powerful theoretical process to a solidly demonstrated one.

Fifty years after Grant, I looked for the finches and was disappointed to have seen only a few, and I had trouble photographing them—they were quick and constantly on the move as they searched for tidbits. However, on my last day in the Galapagos, I was finally successful. Where? In the food court at the Baltra Island airport as I was waiting for my flight. Medium ground finches were hopping around peoples' feet and landing on tables, eating crumbs, and taking bites out of pizzas. While the different species specialize to various degrees, they have not lost the ability to take advantage of an easy resource. The food court was like a wet year when many seeds are available and all finches eat the abundant soft seeds. A sign on the wall said futilely: "Do not feed Darwin's finch."

Brigid at the Regata in Puerto Baquerizo. Mural includes the small ground finch, the medium ground finch, and the large ground finch. Photo by author.

That sign was not the only reference to Darwin in today's Galapagos. He is referenced throughout the islands. Among the guides, they joke that if you don't know an animal's name, claim that is a "lava ___" or a "Galapagos ___" or a "Darwin ___." There is a Darwin Island. The towns are filled with Darwin-named places such as Darwin's Lounge, the bar and restaurant El Muelle de Darwin, Charles Darwin Avenue, and of course the Charles Darwin Research Station.

Although the Galapagos Islands are famous for their unique fauna, and I saw all the notable examples, the density of organisms is greatly reduced from Darwin's day. It is impossible to judge this from a snapshot in time because the populations fluctuate considerably over several years as El Niño and La Niña alternate, creating wet and dry years. But the famous giant tortoises, with their long lifespans, don't experience the short-term fluctuations of most other species. They were decimated by whalers and island residents. The whalers had learned that the tortoises would survive for months, kept upside down without food or water, a source of fresh meat during long months at sea. By the time Darwin arrived in 1835, their populations were greatly reduced.[66] The tortoises are absent from some islands now and rare on others. Various organizations are working to revitalize their populations, with attention to the genetic uniqueness of each island's residents. Ninety-seven percent of the Islands is national park and three percent is devoted to agriculture. Ironically, I saw the greatest number of tortoises in the latter. We visited a farm in the lush agricultural section of Santa Cruz, a farm that sits just outside the national park and is designated as tortoise habitat. We could see as many as six at once, grazing on the tall grasses and herbaceous plants. I felt as though this is what the Galapagos was like originally, and justifies the name. One of the meanings of Galàpagos in Spanish is tortoise.

Santa Cruz has more endemic plants than any of the other islands and supports a vibrant agricultural community. This is because it is one of the older islands, and it has had the time to develop rich soil. In contrast, most other islands are covered in exposed lava, and the vegetation is composed of tough, drought-tolerant plants. We saw lava fields only a hundred years old. Some of the western-most islands have active volcanoes, and all the others have dormant ones. Their conical shapes pierce the horizon in all directions, and with the ubiquitous lava, a primordial feel permeates the place—quite appropriate for its historical significance. Like Hawaii, the Galapagos sits on a hot spot in the Earth's mantle. Moving eastward from the active volcanoes over the hot spot, the islands get progressively older and they sink into the Earth's crust until the oldest ones are undersea mounts. This explains the fact that although the marine and terrestrial iguanas separated at least ten million years ago,[67] the oldest island today is only 4.2 million years old.[68]

The proverbial lack of fear of Galapagos fauna was evident everywhere we went. On first arrival, we had to step around Galapagos sea lions that were lounging around on the pier. On stepping off our panga onto North Seymour, I almost tread on a pair of swallow-tailed gulls that were oblivious to me as they sat on their nest site. And of course, the tortoises paid us no mind. Our rule was to not go within six feet of any animal, but sometimes the animals would break the rule and walk directly towards us. Earlier in this book I noted how, unlike birds, coral reef fishes are unconcerned with my presence. Well, the animals of the Galapagos are like coral reef fishes in this respect. I never found wildlife photography so easy. If you want a closeup, just move closer. Your subject will be unconcerned—except for the lava lizards. They are small enough to be prey to Galapagos hawks and other animals, so they are quite wary.

In a way, this trip was a pilgrimage. I saw the place where my heroes, Darwin and Grant, got their insights. During an earlier pilgrimage to England many years ago, I visited Down House, Darwin's home in Kent, and saw the chair and table where he sat as he wrote *On the Origin of Species*. Theodosius Dobzhanski, the major theorist integrating genetics with evolutionary theory, titled a 1973 essay: "Nothing in Biology Makes Sense Except in the Light of Evolution."[69] It has become clear to me that one theme of this book is about the role of evolution in giving me a framework to understand the living world—and seeing how this belief grew in me.

Acknowledgements

I did not start out to write a book. Brigid Moynahan, my wife, was taking a workshop on memoir writing and encouraged me to join the group. It was run by Rosalind Pace at the Truro Council on Aging. I was encouraged by the supportive responses of Rosalind and the group members. I wrote essays and more essays. My terse modifier-free scientific writing slowly became richer, fuller and more passionate. Eventually, I thought that this collection of essays could be a book. I shared them with Josh Krigman at The Writer's Rock. He made many suggestions, the main one being to consider reorganizing the thing into chapters and generating a book with a clearer thread.

I tried to do that. However, this book would not have seen the light of day had it not been for Crystal Adaway. She pored over my crude early drafts and turned a hodgepodge into a coherent presentation. Her numerous suggestions were always framed with warm support that gave me the desire to forge on. I am grateful for her care.

And Brigid, ever present and supportive. It was while we were driving in the car and I was commenting about my difficulty deciding on a title when she said: "Fieldwork."

Several others commented on parts of the work at various stages. Elizabeth Bradfield helped me develop the Hydrolab chapter into coherent piece. David Kessler read the whole thing and suggested a few additions. My children, Justin Clarke and Serena Clarke-Hanusik, greatly enriched my life. They read some portions and filled in pieces of my lapsed memory, as did Leah Olson, Ed Buskey, Jim Tyler and Ed Brothers. The members of the COA Memoir Group were continually supportive. I heartily thank them all.

Those who were there have remembered things that I have forgotten. And in some instances, they remembered things differently. This book represents the past as it exists in my memory. Undoubtedly the reality was a little different. I hope only slightly so. We will never know.

Endnotes

[1] McLaughlin, E.V. (ed). *Book of Knowledge*. New York: Grolier, 1955

[2] Carson, R. *The Sea Around Us*. New York: Oxford, 1951

[3] Ashbrook, F. *The Red Book of Birds of America*. Racine, Wisconsin: Whitman, 1931

[4] Ashbrook, F. *The Blue Book of Birds of America*. Racine, Wisconsin: Whitman, 1931

[5] Ashbrook, F. *The Yellow Book of Birds of America*. Racine, Wisconsin: Whitman, 1931

[6] Ashbrook, F. *The Green Book of Birds of America*. Racine, Wisconsin: Whitman, 1931

[7] Clarke, R.D. "Postmetamorphic growth rates in a natural population of Fowler's toad, *Bufo woodhousei fowleri.*" *Canadian Journal of Zoology* 52 (1974):1489-98.

[8] Dickens, C. *Great Expectations*. Garden City, NY: Doubleday, 1938

[9] Dickens, C. The Life and Adventures of Nicholas Nickleby. London: MacMillan, 1954

[10] Dickens, C. *The Pickwick Papers*. London: Chapman and Hall, 1837

[11] Strand, H.P. "Fun with a Homemade THERMOPILE." *Popular Mechanics* 119(3) (1963):186-88, 208.

[12] Weiner, J. *The Beak of the Finch*. New York: Knopf, 1994

[13] Clarke, R.D. and P.R. Grant. "An Experimental Study of the Role of Spiders as Predators in a Forest Litter Community." *Ecology* 49 (1968):1152-54.

[14] Gibson, E. J. and R.D. Walk. "The visual cliff." *Scientific American, 202* (4) (1960): 64-71.

[15] https://evelknievel.com/

[16] Sasaki, R. *et al.* "Balancing risk-return decisions by manipulating the mesofrontal circuits in primates." *Science* 383 (2024):55-61.

[17] Robinson, A. "Chemistry's visual origins." *Nature* 465 (2010): 36.

[18] Chen, A. 2018. "'Sea Nomads' May Have Evolved to Be the World's Elite Divers." *Scientific American.* https://www.scientificamerican.com/article/human-sea-nomads-may-have-evolved-to-be-the-worlds-elite-divers/

[19] Tolkien, J.R.R. The Return of the King: The Lord of the Rings, Part 3. New York: Morrow, 2021

[20] de la Haba, L. and M.E. Long. "Belize, the Awakening Land." *National Geographic* 141 (1972):124-46.

[21] Van Mead, N. and J. Blason. 2014. "The 10 world cities with the highest murder rates – in pictures." https://www.theguardian.com/cities/gallery/2014/jun/24/10-world-cities-highest-murder-rates-homicides-in-pictures

[22] de Waal, F. B. M. *Chimpanzee Politics: Power and Sex Among Apes*. Baltimore: Johns Hopkins, 1982

[23] Shakespeare, W. *Richard III*. New York: Penguin, 2017

[24] Diamond, J. The Third Chimpanzee: The Evolution and Future of the Human Animal. New York: HarperCollins, 1992

[25] Blasiola, G.C.Jr. "Quinaldine Sulphate, a New Anaesthetic Formulation for Tropical Marine Fishes." *Journal of Fish Biology* 10 (1977):113-19.

[26] Tinbergen, N. *Social Behavior of Animals*. New York: Springer Dordrecht, 1965

[27] MacArthur, R. H. "Population Ecology of Some Warblers of Northeastern Coniferous Forests." *Ecology* 39 (1958):599- 619.

[28] Whittington, C.M. and C.R. Friesen. 2020. "The Evolution and Physiology of Male Pregnancy in Syngnathid Fishes." *Biological Reviews* 95 (2020):1252-72.

[29] Kraak, S. and J. Videler. "Mate Choice in Aidablennius sphynx (Teleostei, Blenniidae); Females Prefer Nests Containing More Eggs." Behavior 119 (1991): 243-66.

[30] Andersson, M. "Female Choice Selects for Extreme Tail Length in a Widowbird." *Nature* 299 (1982): 818-20.

[31] Clarke, R.D. "Population fluctuation, competition and microhabitat distribution of two species of tube blennies, *Acanthemblemaria* (Teleostei: Chaenopsidae)." *Bulletin of Marine Science* 44 (1989):1174-85.

[32] Clarke, R.D. "Effects of microhabitat and metabolic rate on food intake, growth and fecundity of two competing coral reef fishes." *Coral Reefs* 11 (1992):199-205.

[33] Clarke, R.D. "Diets and metabolic rates of four Caribbean tube blennies, genus *Acanthemblemaria* (Teleostei: Chaenopsidae)." *Bulletin of Marine Science* 65 (1999):185-99.

[34] Oxygen consumption is one way to measure metabolic rate. I used BOD (biological oxygen demand) bottles. These are designed for determining potential oxygen depletion of polluted water. They have

glass stoppers in a tapered neck constructed to accept an electronic oxygen probe. Oxygen is measured at the beginning and end of a set time and the difference is the quantity of oxygen consumed. I did this with a blenny in the bottle and thus measured the oxygen consumed by the blenny.

[35] B. H. and G. A. Bartholomew. "Roles of Endothermy and Size in Inter- and Intraspecific Competition for Elephant Dung in an African Dung Beetle, *Scarabaeus laevistriatus*." Physiological Zoology 52(1979):484-96.

[36] Clarke, R.D., E.J. Buskey and K.C. Marsden. "Effects of water motion and prey behavior on zooplankton capture by two coral reef fishes." *Marine Biology* 146 (2005):1145-55.

[37] Finelli, C.M., R.D. Clarke, H.E. Robinson and E.J. Buskey. 2009. "Water flow controls distribution and feeding behavior of two co-occurring coral reef fishes: I. Field measurements.: *Coral Reefs* 28 (2009):461-73.

[38] Clarke, R.D., C.M. Finelli and E.J. Buskey. 2009. "Water flow controls distribution and feeding behavior of two co-occurring coral reef fishes: II. Laboratory experiments." *Coral Reefs* 28 (2009):475-88.

[39] Stott, R. "The House Where Darwin Lived." *Smithsonian Magazine* (2013) https://www.smithsonianmag.com/travel/the-house-where-darwin-lived-4277158/

[40] Salwiczek, L.H. *et al.* "Adult Cleaner Wrasse Outperform Capuchin Monkeys, Chimpanzees and Orang-utans in a Complex Foraging Task Derived from Cleaner-Client Reef Fish Cooperation." *Plos One* 7(11) (2012): e49068, doi: 10.1371/journal.pone.0049068.

[41] Burnsgalen. "Review of Bravo's 100 Scariest Movie Moments of All Time Movie List." IMDb. July 8, 2021. https://www.imdb.com/list/ls025225335/, 2021

[42] Fiorito, G and P. Scotto. "Observational Learning in *Octopus vulgaris*." *Science* 256 (1992): 545-47.

[43] Mauer, R. "The Real Story behind 'Big Miracle'." *The Anchorage Daily News February 3 2012*. http://www.mcclatchydc.com/2012/02/03/137755/the-real-story-behind-big-miracle.html

[44] Clarke, R.D. and J.C. Tyler. "Differential Space Utilization by Male and Female Spinyhead Blennies, *Acanthemblemaria spinosa* (Teleostei: Chaenopsidae)." *Copeia* 2003 (2003): 241-47.

[45] Lidz, F. "What to Do with a Bug Named Hitler?" *New York Times December 27, 2023*. https://www.nytimes.com/2023/12/26/science/taxonomy-beetle-insects-hitler.html

[46] St. George, Z. "When Species Names Are Offensive, Should they Be Changed?" *Yale Environment 360,* January 4, 2024. https://e360.yale.edu/features/renaming-species-offensive-names-taxonomy-nomenclature

[47] Williams.J.T, and J.C. Tyler. "Revision of the western Atlantic clingfishes of the genus *Tomicodon* (Gobieosidae), with descriptions of five new species." *Smithsonian Contributions to Zoology* Number 621 (2003).

[48] Wohlleben, P. and J. Billinghurst (Translator). *The Secret Wisdom of Nature: Trees, Animals, and the Extraordinary Balance of All Living Things*. Vancouver: Greystone, 2019

[49] Naylor, R. "What is a vortex?" https://visitsedona.com/spiritual-wellness/what-is-a-vortex/

[50] Jackson, J.B.C. *et al*. "Historical Overfishing and the Recent Collapse of Coastal Ecosystems." *Science* 293 (2001): 629-37.

[51] Kaufman, L.S. "Threespot Damselfish: Effects of Benthic Biota of Caribbean Coral Reefs." *Proceedings, Third International Coral Reef Symposium, University of* Miami (1977): 559-64.

[52] Clarke, R.D. "Population shifts in two competing fish species on a degrading coral reef." *Marine Ecology Progress Series* 137(1996): 51-58.

[53] Møller, A.P. "Parallel declines in abundance of insects and insectivorous birds in Denmark over 22 years." *Ecology and Evolution* 9 (2019): 6581-87.

[54] Carson, R. *Silent Spring*. Cambridge, MA: Riverside Press, 1962

[55] Leopold, A. 1949. *A Sand County Almanac*. New York: Oxford, 1949

[56] Svetlana A. C. *et al*. "Materially engineered artificial pollinators." *Chem* 2 (2017): 224-39.

[57] Young, J.Z. *The Life of Vertebrates, Second Edition*, Oxford: Oxford University Press, 1962

[58] Dobzhansky, T. *Genetics and the Origin of Species*. New York: Columbia Univ. Press, 1937

[59] Wilson, E.O. *Sociobiology: The New Synthesis*. Cambridge: Belknap, 1975

[60] Olson, R. *Don't Be Such a Scientist*. Washington: Island Press, 2009

[61] Dawkins, R. *The Blind Watchmaker*. New York: Norton, 1986

[62] Yong, E. An Immense World. New York: Random House, 2022

[63] Oliver, M. *Blue Pastures*. New York: Harcourt Brace & Co, 1995

[64] De Roy, T. *A Pocket Guide to Birds of Galápagos.* Princeton: Princeton University Press. 2022

[65] Lamichhaney *et al*. "Rapid hybrid speciation in Darwin's finches." *Science* 359 (2017):224-28

[66] Bercaw Edwards, M.K. "Of Melville, Tortoises, and the Galapagos." *Historic Nantucket* 64 (Fall 2014):12-15.

[67] Rassmann K. "Evolutionary age of the Galápagos iguanas predates the age of the present Galápagos islands." *Molecular Phylogenetics and Evolution* (1997): 7(2):158-72.

[68] Volcanic Galapagos. https://darkwing.uoregon.edu/~drt/Research/Volcanic%20Galapagos/index.jsp.html

[69] Dobzhansky, T. "Nothing in Biology Makes Sense Except in the Light of Evolution." *American Biology Teacher* 35 (1973): 125–29

www.ingramcontent.com/pod-product-compliance
Lightning Source LLC
Chambersburg PA
CBHW051614010526
44107CB00036B/1427/J